AAPS Advances in the Pharmaceutical Sciences Series

Volume 27

The AAPS Advances in the Pharmaceutical Sciences Series, published in partnership with the American Association of Pharmaceutical Scientists, is designed to deliver volumes authored by opinion leaders and authorities from around the globe, addressing innovations in drug research and development, and best practice for scientists and industry professionals in the pharma and biotech industries. Indexed in

Reaxys

SCOPUS Chemical Abstracts Service (CAS) SCImago

EMBASE

Robert R. Nesbitt
Ramakrishna Venugopalan • Wolfgang Fraunhofer
Editors

Patient-focused Development of Biopharmaceutical Products

Springer

aaps®

Editors
Robert R. Nesbitt
AbbVie, Inc.
Gurnee, IL, USA

Ramakrishna Venugopalan
Continuity Biosciences LLC
Malvern, PA, USA

Wolfgang Fraunhofer
Drug Product Development
AbbVie, Inc.
North Chicago, IL, USA

ISSN 2210-7371 ISSN 2210-738X (electronic)
AAPS Advances in the Pharmaceutical Sciences Series
ISBN 978-3-031-98252-1 ISBN 978-3-031-98253-8 (eBook)
https://doi.org/10.1007/978-3-031-98253-8

AbbVie Inc., 1 N Waukegan Rd, Chicago, IL 60601, USA

This Springer imprint is published by the registered company Springer Nature Switzerland AG
The registered company address is: Gewerbestrasse 11, 6330 Cham, Switzerland

Preface

To the reader.

Thank you for your interest in patient focus and your willingness to learn more about the ultimate consumer of any drug/device/combination product or therapy—the patient.

In this book, we will share what the collection of industry experts have learned about how to understand the needs of the patient, and how to keep that front and center through the entire product definition, development, regulatory approval, launch, and full product life cycle of any such therapy. This book is the compilation of input from an industry-wide team of contributing authors, all passionate about better understanding of patients' needs and keeping those needs in mind throughout the development and support of these products.

Unlike some texts where individual chapters can be sourced individually, we have written this book—a collection of learnings pulled from years of experience across a wide range of disciplines and therapies—as a unified story which we intend to weave learnings and best practices as well as considerations to manage the seemingly conflicting demands which developers of biopharmaceutical products will experience.

In **Chap. 1**, we share a background in the research and development practices, applicable to any product development activity—but especially important in grounding the crucial early understanding of the user's needs and how those can be transitioned into requirements and ultimately a product concept. The key concept is that innovation in any space—but certainly in developing a therapeutic treatment—starts with understanding the user's (the patients') needs. We start with methods to conduct this initial and crucial discovery in which those needs (often unspoken) can be understood, where qualitative learnings can provide insights to those unmet needs. When well conducted, this early research can evoke a better understanding of the patients' journey through their disease, especially chronic diseases. This chapter provides a basis for Chap. 2, in which we broaden our focus on the patient and their needs, and work to understand the world in which they live.

Chapter 2 takes off from the patients' needs and seeks to understand the "ecosystems" in which patients live. This differs by geography, individual circumstances, and standards of care. Understanding the context in which the patient needs to treat their disease is vital to ensure that any product we develop makes sense in the "real world" in which they live. In this chapter, we share a couple of different vignettes—examples of people dealing with their diseases—not just what they use to treat their disease or manage their symptoms but how the world around them influences (and sometimes challenges) their ability to deal most seamlessly with the intended therapy. Examples of research opportunities to better understand the patient ecosystems are provided throughout the vignettes to inform specific contexts of use and understand usability issues you may encounter with your specific patients.

Then, based on what we have seen so far—how to understand the users' needs and develop concepts to address them, and how to understand the world in which patients need to make those solutions (when these concepts would be realized)—we are prepared to identify the specific patient needs which form the product concept to be developed. Thus, **Chap. 3** looks at defining the User needs, and the considerations in determining the key criteria by which the ultimate product concept will be measured. This includes the dosing (how much and at what frequency), the formulation (how the drug product is mixed and presented) as well as the key characteristics of the presentation (size, features, how it is used, etc.). The chapter addresses the concepts of delivery platforms—common or consistent types of devices or combination product presentations, which can be deployed for different therapeutics. This chapter also illustrates the drug development process, how clinical trials are used to confirm the effectiveness of the drug product, and the product presentation in which that product is deployed (e.g., combination product). This chapter also addresses considerations for populations with special needs, such as pediatric users. This chapter also looks at the considerations that need to be balanced—involving the patients and their capabilities, the capabilities of the device or delivery system. Finally, this chapter looks at how to understand and incorporate the growing need to incorporate connectivity into the therapeutic experience (above and beyond the actual drug delivery).

By this point in the book, we have seen techniques to understand the user's needs, how product concepts are created, how those concepts are tested with users, how clinical trials and designed, and how to address the array demands that must be met to help this therapy fit into the "real life" in which patients live. However, the real challenge a biopharmaceutical therapy developer must address is how to balance the tradeoffs to address the time and needs of this moment. **Chapter 4** looks at the various criteria that the developer must balance to create the right solution at the right time. Engaging a cross-functional team throughout the full process—from initial needs assessment, concept creation, all the way through development, clinical trials, regulatory preparation and throughout product launch and support of the launched product—is key. The Commercial team—including the Marketing organization (by whatever name it is called)—which translates the high-level business strategy into disease state strategies and business needs, which drive the "problem to be solved"—must not just be involved but "own" this process. Which leads us to **Chap. 5** in which we explore the Commercial viewpoint.

In **Chap. 5**, we go beyond the traditional role of product positioning and launch and explore how the Commercial team must be engaged much earlier in a very strategic fashion—and own and steer the full product life cycle. This chapter looks at how to understand the target patient—their mindset, and what drives (or hinders) their use of (or failure to use) products (drugs, combination products, delivery systems) in this space treating the disease. The Commercial (or Marketing) team needs to be involved early and throughout the development, from the earliest discovery research (as described in Chaps. 1 and 2), as well as through the full development path and then continue to own and steer the product as it matures in the market (and faces newer competitive offerings). In this chapter, we look at patient needs and motivators in different disease states and product presentations (e.g., testosterone treatment via underarm topical administration, and intravenous (IV) administration in a clinical setting). The chapter also walks through a case study for a patient self-injection drug product and the decision-making in planning an additional (third) product presentation (delivery system) for a therapy. Finally, this chapter wraps up with some key high-level observations or learnings gathered from the authors' collective industry experience (a "top ten" list if you will).

In **Chap. 6**, we discuss the regulatory viewpoint as it relates to planning and executing a product presentation for a patient-focused therapy. However, the fact that this chapter is chronologically last should not imply that an understanding of the regulatory climate and the needs and expectations of the regulatory agencies in the target geographies are to be left to "the final steps." Rather, an understanding of the real regulatory climate and expectations must be incorporated throughout—from the earliest concept development stage. The cross-functional product design and development team must include a regulatory "owner," who can help influence the world in which such a product once envisioned and developed will exist. This chapter explores the examples discussed in this book and explores regulatory considerations and strategies.

As a way to demonstrate how patient needs have been addressed through the years, **Chap. 7** presents a case study that explores the 100-plus year history of the treatment of diabetes. This is understandably a high-level overview of how patient needs have been addressed in the evolution of therapies (including diagnostics, therapeutics, and delivery systems) which have been available for persons with diabetes over that time.

Finally, in **Chapter 8**, the authors of the book seek to share some of the key learnings and areas of particular focus, which we would suggest you, the reader, take into your work in the development of drug delivery systems and medical devices.

We, the editors and authors of this book, applaud your interest in understanding the needs of the patients that motivate our work and drive the impact we all choose to make in the world.

Enjoy!

Gurnee, IL, USA — Robert R. Nesbitt
Malvern, PA, USA — Ramakrishna Venugopalan
North Chicago, IL, USA — Wolfgang Fraunhofer

Contents

About the Editors

Robert R. Nesbitt is Senior Director, Landscaping & Platform Definition in AbbVie's Combination Product Development group. Prior to that, he was Director, Human-Centered Design & Human Factors, and led the team that drove user needs assessment and usability validation for AbbVie Combination Products. He was co-editor of *Development of Biopharmaceutical Drug-Device Products*, a book for pharmaceutical development professionals on drug product formulation, development and the delivery systems, which are required to deliver them. This 800-page book was published by Springer as Volume 35 of AAPS' Advances in the Pharmaceutical Sciences series (2020).

He joined AbbVie in January 2015, after 16 years at Eli Lilly and Company, where as Director of Advanced Development and Research, he built patient-centered design into early concept design and development of new Lilly delivery devices. His work drove the design of nine different Lilly delivery devices in use today globally—his team led the design of the device Lilly uses for Trulicity, Emgality, Taltz, Mounjaro, and Zepbound. After retirement from Lilly (2015), he was called back and recognized as a "Top Innovator" for his leadership in driving the patient-centered design of the auto-injector used for three Lilly therapeutics.

Prior to Lilly, he led process engineering and manufacturing for Roche Diagnostics' blood glucose monitoring and Baxter Healthcare's custom sterile operating room procedure pack businesses.

Active in the development of ISO standards for medical devices since 1999, he is an Expert on multiple standards groups, was Convener for several Working Groups, and was ISO TC84 Chair from 2022 to 2024. He was awarded the ISO Excellence Award for years of service to ISO TC84, and leading the development of ISO 11608-7, "Needle-based injection systems - Accessibility for persons with visual impairment" (2017). He has a BS in Engineering Operations from Iowa State University and an MBA from the University of Chicago.

Ramakrishna (Krishna) Venugopalan is currently the Co-Founder and CEO of Continuity Biosciences, LLC, a biotechnology company established with the mission of revolutionizing therapeutic delivery. At Continuity, proprietary platform technologies are leveraged to develop next-generation drug delivery systems aimed at addressing bioavailability challenges and mitigating systemic toxicity, particularly in oncology and metabolic diseases.

Prior to founding Continuity, Krishna led AbbVie's combination product, drug delivery, and sterile packaging development organization, integrating medical technology with biotechnology within AbbVie's pipeline. He played a central role in the launch of impactful therapies, including Skyrizi®—an IL-23 inhibitor widely used to treat autoimmune diseases, and Vyalev™—a continuous subcutaneous infusion therapy providing enhanced symptom control for patients with advanced Parkinson's disease.

Before his tenure at AbbVie, he held leadership positions at Johnson & Johnson focused on drug-device combination development, M&A integration, and external partnerships. In 2016, he received the Johnson Medal for co-developing a continuous glucose monitor integrated insulin pump, an effort that complemented his research and development focus on closed-loop insulin delivery systems for type 1 diabetes. He also led initiatives in neuromodulation, implantable drug delivery, and intracranial fluid management, earning multiple Standards of Leadership Awards recognizing his impact across both R&D and commercial domains.

Earlier in his career, Krishna served as a faculty member in biomedical engineering at the University of Alabama at Birmingham (UAB), where he established a translational research laboratory focused on minimally invasive medical devices. His collaborations with over 20 industry partners facilitated the advancement of more than a dozen products into clinical trials or commercial use. This work contributed to the development of an ASTM standard and an FDA guidance document on stent reliability.

Krishna holds a BE in Electronics and Instrumentation from Annamalai University, a PhD in Biomedical Engineering from the University of Alabama at Birmingham, and an MBA from MIT's Sloan School of Management.

Wolfgang Fraunhofer is Senior Director, CMC Strategy and Portfolio Leadership, in Development Sciences, R&D at AbbVie. Previously, he led teams in NCE Analytical R&D, Combination Product Development, NBE Formulation Development, and NCE/NBE Pharmaceutics/Preformulation.

He has deep academic and professional experience in pharmaceutics and drug product development. He obtained a Pharmacy degree from Universität Regensburg in Germany, followed by a Doctorate in Pharmaceutical Technology and Biopharmaceutics from Ludwig-Maximilians-Universität München, Germany. He continued to broaden his expertise with a Master's Degree in Biotechnology from the University of Wisconsin-Madison, USA.

He began his pharmaceutical development career at BASF/Knoll, first while completing his doctoral research on protein aggregate characterization, and later as member of the team developing formulations and product presentations for adalimumab, one of the most widely used monoclonal antibodies that has transformed immunology treatments. His pharmaceutical expertise has been deployed in the development of various pharmaceutical excipients as well as the development of lyophilized and liquid new biological entities (NBEs). Upon Abbott Labs purchase of BASF's Knoll Division, he worked on Drug Product scale up and transfer to manufacturing sites at locations in Germany as well as Puerto Rico, supporting the development of manufacturing quality criteria and training of the operators in the manufacture and final GMP release of adalimumab (Humira®). He led the development of selected scientific strategies as well as the creation of relevant intellectual property (IP) positions in this space.

He led teams in Pharmaceutics to support preclinical pharmacokinetics, efficacy and toxicology studies, physico-chemical and stability characterization of new chemical entities (NCEs) as well as NBEs, including new modality formats like dual-variable-domain Igs (DVD-Igs™). His work has impacted many assets across all phases of clinical development, including global NDA and BLA submissions of various NCE and NBE assets in Immunology, Oncology and Women's Health. In addition, he led teams of engineers in the development of medical devices and combination products to support Abbott's and AbbVie's drug pipeline.

As Senior Director of CMC Strategy and Portfolio Leadership at AbbVie Inc., he and his team of CMC Product Development Directors drive CMC and Product Presentation cross-functional team as well as processes of CMC Development from GLP tox through clinical development, regulatory submissions, and launch support. His academic and cross-functional experience in all phases of pharmaceutical development across modalities and various Therapeutic Areas make him a leader in enabling and differentiating AbbVie's Pharma Pipeline. Throughout his career, he has been focused on understanding patient needs and translating evolving science and MoAs into life-changing and competitive treatment options.

Chapter 1
Leveraging Discovery Research to Inform Product Design

Steve McPhilliamy and Carolyn Rose

Abstract This chapter introduces the human-centered design approach and the importance of understanding the patient experience. This chapter will delve into the role of discovery research to serve as a foundation for decision-making and design, ensuring that new products meet the needs and expectations of those who will use them. It outlines the early-stage process and key methods to inform drug and delivery device development, from applied ethnography to drafting concept visions, and provides real-world anecdotes and examples throughout to demonstrate practices and learnings. Lastly, the authors provide tips and tricks for success: succinct guidelines and common pitfalls learned over the authors' collective 40+ years in the field.

Keywords Human-centered design · Discovery research · Ethnography · Patient journey

Introduction

This chapter delves into the significance of early-stage research that utilizes a human-centered design approach. The basis of any successful product solution – whether it is a diagnostic app, a surgical tool, or a combination product – the foundation starts here. By understanding behaviors, attitudes, expectations, and needs of the patient in culture and context, it creates the groundwork for meaningful innovation (Fig. 1.1).

For over 30 years, researchers in health care have been conducting human-centered design. They have spent countless hours in hospitals and patients' homes,

S. McPhilliamy (✉)
Design Stack LLC, Chicago, IL, USA
e-mail: smcphilliamy@design-stack.net

C. Rose (✉)
Design Research, Tangible LLC, Chicago, IL, USA
e-mail: Carolyn@tangiblefindings.com

R. R. Nesbitt et al. (eds.), *Patient-focused Development of Biopharmaceutical Products*, AAPS Advances in the Pharmaceutical Sciences Series 27,
https://doi.org/10.1007/978-3-031-98253-8_1

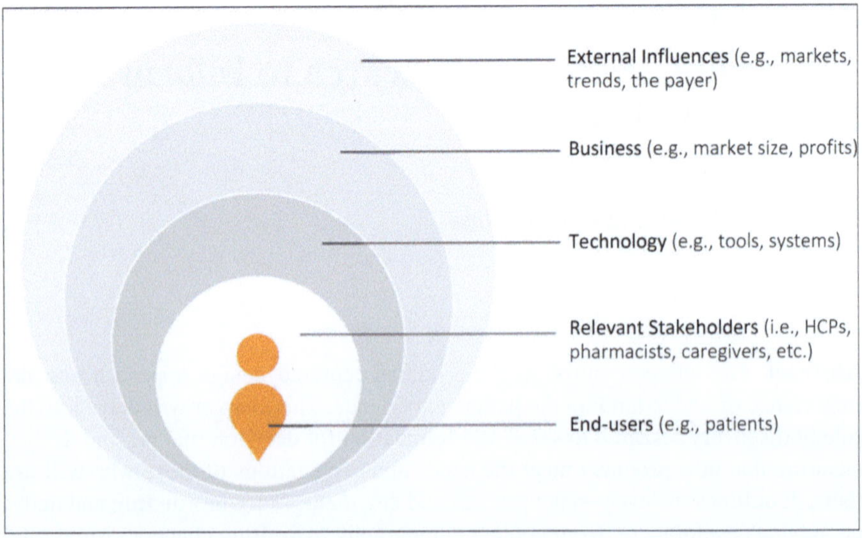

Fig. 1.1 A human-centered design approach

interviewing, observing, and documenting findings. In this chapter, we take a deep dive into the patient journey and how it plays a crucial role in defining the entire user experience. We will share thoughts on the team needed to define the focus and to implement the solution. We will describe our research output and deliverables and will discuss how to keep patients at the core of design and development. In the book, you will see how each chapter will build upon one another and "connect the dots" from early needs finding to later stage commercialization and all the steps between.

First, a note about the people and backgrounds needed to conduct great research. Empathy, compassion, agility, and a strong curiosity are critical elements that researchers must possess. Researchers spend a great deal of time with people, observing their lives, struggles, and pains in real time. They witness personal interactions, elements that inspire, and moments of extreme vulnerability. They must dynamically relate, adapt, and learn with the hopes of linking unmet needs with innovation opportunities.

> You never really understand a person until you consider things from his point of view.—Harper Lee[1] (from To Kill a Mockingbird, Lee, Harper, Philadelphia, Lippincott, 1960)

What Is Discovery Research?

Imagine you're a detective, tasked with solving a mystery. In addition to sitting behind a desk, poring over data and numbers, you're out in the field, talking to people and observing their behavior in their natural habitat. That's what discovery research is all about.

It's a type of investigation that's grounded in the social sciences, and it relies on ethnography—the study of people and cultures through immersion. By immersing yourself in the real-life situations experienced by others—family interactions, time pressures, uncertainties, treatment routines—you gain a deeper understanding of their needs and what drives them.

Qualitative, Not Quantitative

Unlike larger-scale market research, which often focuses on quantitative data and large statistically relevant sample sizes, discovery research is qualitative in nature. It's about going deep, not wide. This means the research sample will be smaller, but the insights will be richer (Fig. 1.2).

You might be wondering: how do we choose the right sample size for this research? Well, it's a delicate balance. You want to have enough people to see patterns and trends, but not so many that the data becomes overwhelming and time consuming. In our experience, a group as small as six can start to reveal patterns, provided they share key characteristics that define their experience.

For example, let's say we're conducting a study on the patient experience surrounding diabetes management. We might include six people with Type 1 diabetes, six people with Type 2 diabetes, eight prescribers, and four diabetes nurse educators. Within each patient group, we'd seek a range of experience levels, symptom

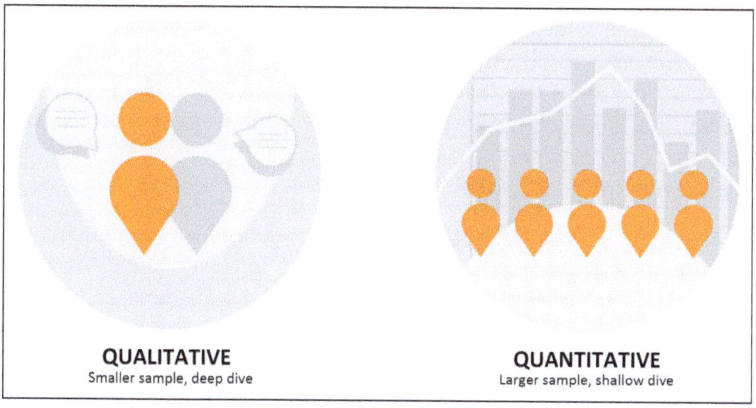

Fig. 1.2 Qualitative vs. quantitative research

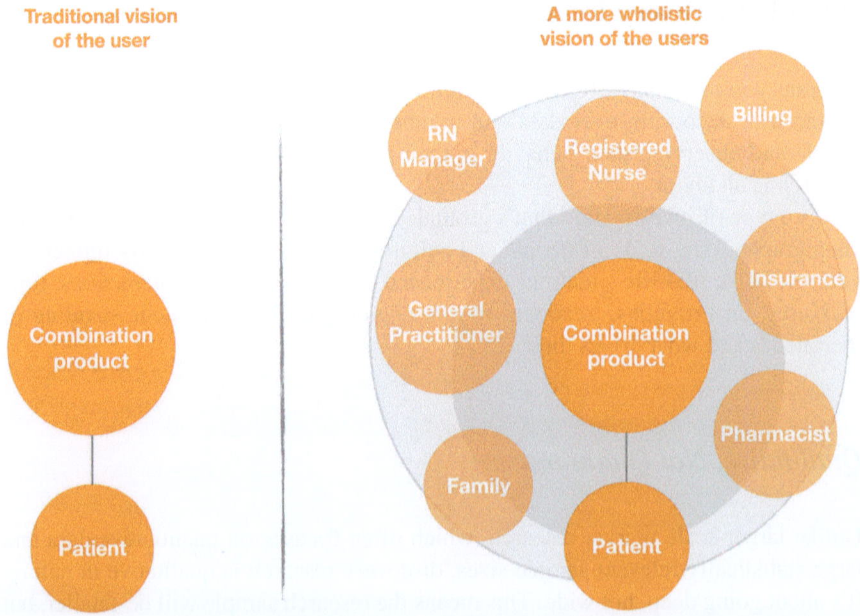

Fig. 1.3 Mapping relevant stakeholders and roles can help inform research sample

severity, and treatment modalities. By considering both patients as well as those who influence the patient experience (HCPs, caregivers, etc.), we can gain a more comprehensive understanding of the patient journey and the needs of different stakeholders (Fig. 1.3).

Discovery Methods

Discovery research is about getting out into the world, listening to people, and observing their behavior. It's a qualitative approach that yields rich insights and a deeper understanding of the "why" behind behaviors, emotions, perceptions, and preferences. For those who truly want to uncover the truth around behaviors and insights, there's no better way to do it.

The discovery methods used include:

- Narrative tours—Here we ask people to act as a tour guide throughout any space relevant to a particular experience or practice of interest. Patients may highlight where and how treatments and supplies are stored, for example, calling out rationale as well as likes and dislikes associated with the space.

Recently, during a narrative tour, we learned that an RA patient keeps her sharps bin on the upper shelf of a hallway closet – nowhere near her autoinjector or other supplies (all co-located in her bedroom). She commented here it's "out of sight" from guests and "out of reach" from her children, but also serves as a prompt for her to manage (i.e., drop off and replace) as needed.

- Contextual observation—This technique is leveraged extensively for discovery research, where we are often "flies on the wall" observing a particular activity or interaction. This typically focuses on particular activities of interest, for example, one's treatment routine, prescription pick-up, a patient consultation at the doctor's office, etc.
- Shadowing—This involves following someone around throughout their day (ideally without disruption!). This is an observation method similar to the above but is comparatively more dynamic and less focused. For example, we might shadow a clinician throughout their workday to get a feel for their workflow, practices, and patient interaction.
- Video diaries—This is a remote and often asynchronous observation method (rather than real time) that allows us to target very specific and perhaps spontaneous moments often over longer durations of time. This is a great tool for better understanding patterns and variations in behavior and the things, which influence that behavior (e.g., environment, time of day, activity, etc.).

In a study aimed to better understand the patient experience surrounding glaucoma (and dry eye, which often results from glaucoma treatment), video diary entries allowed us to observe how patients administer treatment throughout their day – including at home, work, in the car, and even at the grocery store. While the task of administering an eye drop remained the same, the approach and challenges varied significantly based on one's surroundings (both environmental and social) introducing unique needs compared to at-home administration.

- One-on-one interviews—When possible, we always recommend that interviews be conducted one-on-one (rather than in groups) in order to have a more direct, deep, and open conversation. We do find value in meeting with patients' family/caregivers (it's not uncommon for a partner or child to chime in with "the real truth" around someone's dietary habits, treatment adherence, and recent lab tests—which we may or may not have been inquiring about!), but even still we prefer to structure the interview such that each contributor also participates without the presence of the other(s).
- Storytelling—Because the nature of this research doesn't allow us to observe every scenario or nuance, and as the saying goes "you don't know what you don't know," we rely on storytelling to fill in the gaps. For example, asking patients to share a personal disaster/success story related to treatment management, or a time they felt proud, confident, uncertain, or scared allows us to broaden our lens.

An asthma patient well versed in the use of her current inhaler later told us that she misused her previous metered dose inhaler unknowingly for roughly four years until an exacerbation leading to hospitalization (her "scary moment") prompted her and her care team to re-evaluate treatment efficacy and use.

- Hypotheticals—As design-minded people, teams often have hypotheses even before fieldwork and early ideas start to germinate with the first observed challenge. While some suppress this inclination, we encourage it, and start to test these inklings in real time. Often at the conclusion of a research visit (so as to not bias the participant-directed activity and discussion), we may ask, "What if…?" to gauge response and begin to shape a future state.
- Card sorting—This method is grounded in co-creation, inviting research participants to engage in the design process itself. Card sorting can be considered an early co-creation method that we often leverage to get a sense for relative value and/or importance, often of experiential attributes (e.g., safety, confidence, ease of use, etc.) and/or feature sets (e.g., adjustable needle depth, LED feedback, reminders, etc.).

These methods provide unique lenses through which to better understand research participants' daily lives and interactions, as well as the things that motivate them. By using a variety of techniques, the research team can gain a more comprehensive understanding of the experiences and desires of the participants and use that information to inform future designs and solutions. Understanding goals allows us to identify transformative opportunities that align with user needs and increase impact.

What Happens Next?

Discovery research yields lots of data—it can be a daunting task to try to make sense of. Yet, data synthesis is arguably the most important aspect of the process because without it the data is not actionable and is unlikely to foster alignment. Below we have outline, loosely in chronological order, what we view to be four key frameworks/outputs we utilize both to synthesize data and translate it into early design and development efforts.

The Patient Journey

One of the first tools that is utilized in this synthesis is the development of a "patient journey"—a pictorial representation of the steps that the patient (and often family members and/or health-care professionals) goes through in the evolution of their disease. An overview of such a patient journey is shown in Fig. 1.4.

We approach the patient journey as a distilled visual that deconstructs the patient experience, from diagnosis through ongoing disease management, into what proves

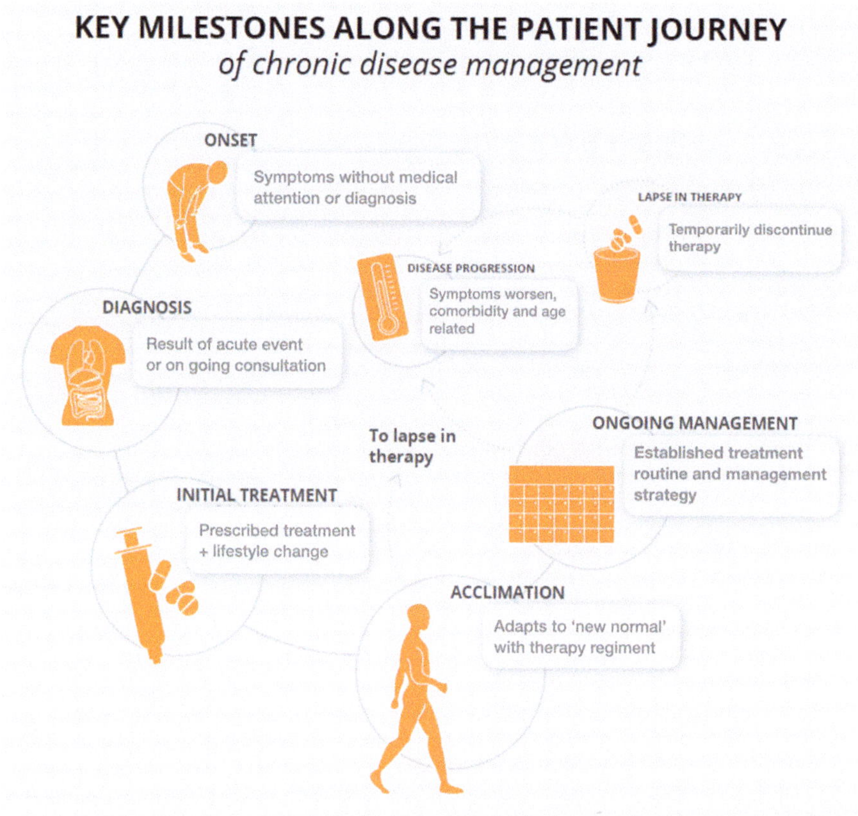

Fig. 1.4 The patient journey maps pivotal moments from diagnosis through ongoing disease management

to be pivotal moments. This visual serves to make digestible what can be a life-long journey, calling out key stakeholders, roles, relationships, actions, and decisions as well as the impetus and impact of each. Within the patient journey, we may also zoom into a "day in the life" of a patient, so as to understand their current treatment regimen—along with any challenges, frustrations, and variations experienced, as well as the drivers and implications of each. (You will see some examples of this through the use of vignettes in Chap. 2.)

Opportunity Framing

Over time we have landed on a formula of sorts for how to best frame an opportunity. Ideally, opportunities are goal-oriented and solution agnostic, and stand on their own (almost the "*Cliff Notes*" version and/or punchline of what can be 100+ slides of findings).

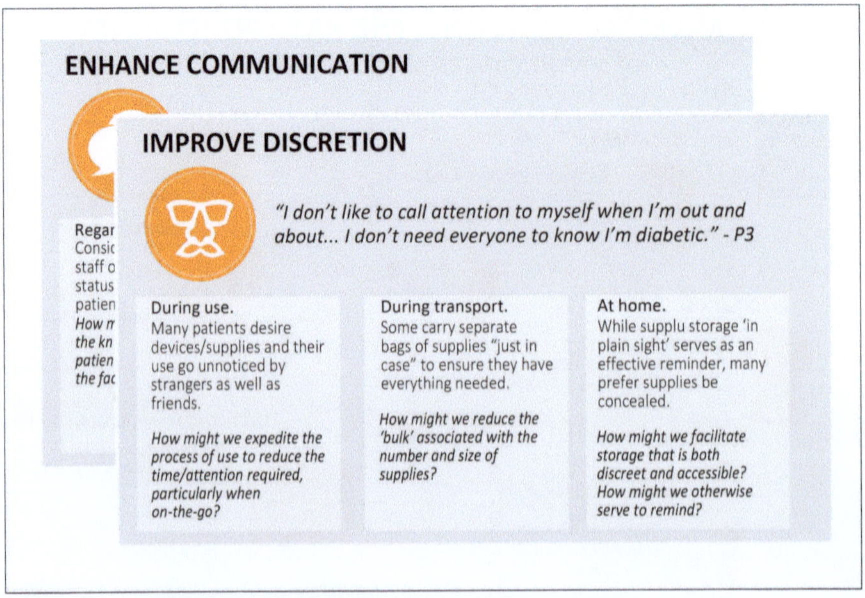

Fig. 1.5 Sample opportunities

Each opportunity we draft includes

- The high-level goal (written in the users' voice).
- The context (meant to communicate its intended value and purpose, e.g., current challenges, ideals, etc.)
- The "evidence" (e.g., quotes, photos, artifacts from the field).
- Directives (a compilation of considerations that may direct solutions).
- HMW questions (e.g., how might we ensure a full dose is consistently received?) (Fig. 1.5)

Now may be a good time to mention the important distinction between goals and tasks. The tasks that we observe someone carry out often are a means to an end; if we seek to optimize tasks in the absence of understanding higher-order goals, any and all opportunities will be incremental at best.

Concept Visions

During the early design process, we generate and organize ideas based on the identified opportunities. We group and build on complementary ideas—what we consider to be key attributes and enablers—to create a newly envisioned experience. A concept vision includes a storyboard that depicts this envisioned experience from the users'/patients' perspective. Figure 1.6 shows a sample concept vision, which

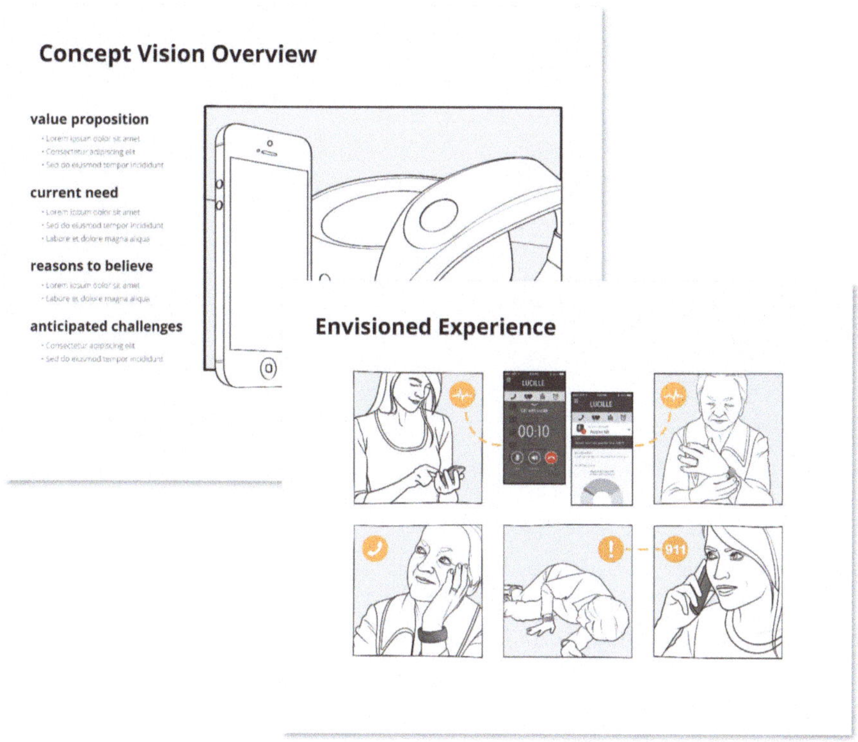

Fig. 1.6 Sample concept vision

includes an illustration of the potential solution (on the left) and the user experience enabled by this solution (on the right).

These storyboards prove to be an effective and low investment means to communicate early concepts for feedback. We quickly put these concepts in front of potential end users to understand the extent to which they deliver on the intended value, and learn of any perceived barriers, concerns, or challenges. This feedback allows us to both quickly narrow to the concept or set of concepts likely to yield the most value and inform design modifications/refinement.

It's critical, of course, that concept vision creation and evaluation be a collaborative activity. While we don't necessarily want to invest the R&D dollars (yet!) to make this real, we do want to ensure we have "reason to believe" concepts are viable from a technical, market, and regulatory perspective. We like to integrate this context both before and after end-user feedback, the former to keep us from sharing a vision that can't be realized, the latter to accompany end-user feedback as an input for down-selection and refinement.

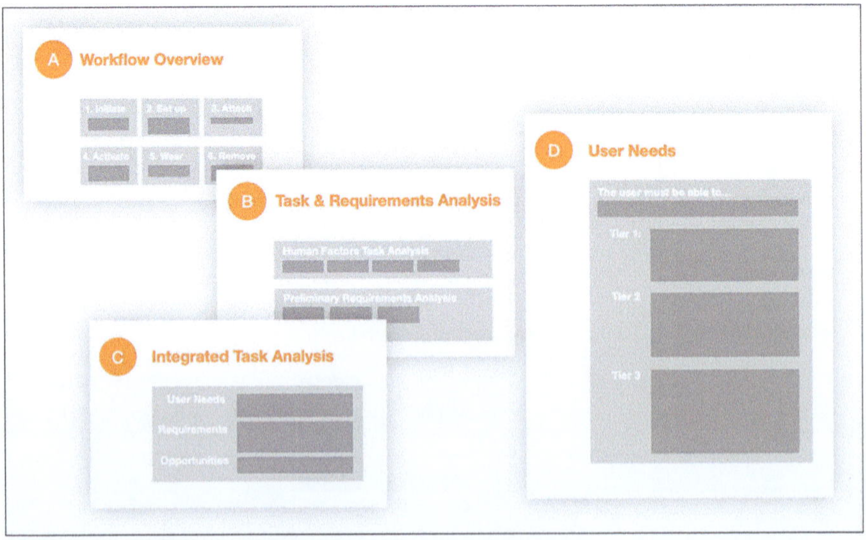

Fig. 1.7 Components of a concept development framework

Our Concept Development Framework

Once a concept has been defined (in terms of value proposition, key attributes, and enablers, as well as envisioned experience), a critical next step in our process is the development of a concept development framework.

This framework is meant to be the "action plan" for design and development activities, integrating user needs and design opportunities with technical, marketing/business, and regulatory inputs in a prioritized matrix. Over the years, we repeatedly (and at times unsuccessfully) navigated the conflict of "locking down" requirements during early-stage development while also allowing for flexibility integral to design best practices. This conflict can lead to wasted efforts exploring something that is later deemed unviable and late changes/updates that impact prior work in unexpected ways. The concept development framework is meant to address this tension by highlighting both design opportunities and fixed constraints and allowing for alignment (and debate!) prior to any development activity (Fig. 1.7).

Tips and Tricks for Success

Challenge Assumptions

...Both your own and your organizations assumptions are likely to exist and can easily go unrecognized (or even be disguised as truth!). Discovery research should be leveraged to confirm or refute any assumptions that have prevailed across the organization over time. Throughout observation and interviews, always ask why even when you think you know the answer. (The response may surprise you.)

Looking at supplemental oxygen use (stationary and portable systems), one might assume using a smaller, more portable system within the home would allow you more mobility. However, in conducting research with COPD patients, we came to learn that portable oxygen systems are often perceived to be more of a hindrance for home use. While larger, stationary oxygen requires tubing follow you throughout the home, most preferred this to having something over their shoulder and/or in close proximity at all times. Interestingly, current portable systems, while praised for enabling "freedom" that equated to activities from social events to golf outings, were perceived to be burdensome within the confines of one's home.

Think Holistically

The patient journey is just that—a collection of moments and experiences over time that together comprise an experience. As such, it's important to be mindful of the entirety of the journey, recognizing that the whole is often greater than the sum of its parts, and experiences are shaped by many aspects likely to occur before and after (particularly if an on-going or routine experience such as a treatment regimen).

Within pharma, there understandably tends to be a focus on the administration event itself, often "starting" with acquiring supplies and "ending" with disposal. However, in a recent study conducted with people treating hemophilia (both patients and caregivers), a number of unique challenges were identified surrounding onboarding, growing with treatment (managing new dosing regimens over time), and the 'in-between' – that which happens between doses, including storage and replenishment, leading to design opportunities from package differentiation (particularly since multiple family members are likely to receive treatment) to expiration management.

Prioritize

The prioritization of what we call experience drivers (thematic influencers of a user experience) allows a development team to effectively navigate inevitable trade-offs. (If the user wants a large display on a small, portable device, now what?)

For people with both Type 1 and Type 2 diabetes, confidence ("did I get my dose?") and discretion ("did it go unnoticed?") are typically important drivers of the patients' experience. However, for those taking short-acting (i.e., bolus) insulin throughout the day (more commonly those with Type 1 diabetes) – confidence can be less important than discretion comparatively due to both experience level and context of use (often out in public). For the same reasons, the relative importance of the two drivers may be inverted for those taking long-acting (i.e., basal) insulin once daily before sleep. If we consider the impact this prioritization may have on device feedback, for example, those taking basal insulin may prefer more overt modes of feedback directly on the pen, while those taking bolus may prefer more subtle modes of feedback and/or indirect (e.g., confirmation via app).

Iterate

Design iteration is critical to success. We introduce concepts to current and prospective end users early and often, increasing the concept fidelity with each round of testing until we feel confident it meets the needs and expectations of each stakeholder involved. Each time we do so, we encourage research participants to compare concepts to current practice—whatever it might be—to understand perceived benefits and shortcomings, both to strengthen the value proposition and address any user-driven challenges/concerns.

> A longitudinal study on chronic pain management identified an opportunity to better objectify pain such that patients could more readily assess the impact of treatments (typically largely based on the abstract numeric pain scores that often have little meaning to a patient's day-to-day). From this, we developed a patient-facing app to assess pain, which ultimately narrowed to a set of three daily quality of life (or QOL) questions that pain typically impacts (e.g., activity level, sleep, mood,) empowering patients to assess their chronic pain more readily over time and identify patterns (i.e., trends mapped to treatment).

A Note on Digital

With the abundance of apps and an ever-increasing focus on digital solutions and the role they may play in empowering patients and promoting adherence, it's critical that we take the time to understand the user experience and actual needs. We caution with the familiar saying "just because you can, doesn't mean you should," as we have seen countless iterations overwhelm and frustrate end users who ultimately decide digital engagement "isn't worth it." Capturing data in a way that's meaningful and motivating to patients is no easy feat, nor does it encourage ongoing interaction. It requires an in-depth understanding of user behavior, attitudes, and motivations to create an experience that truly supports patient needs.

For more on the use of connected health or digital solutions in health care, refer to Chap. 3, where there is a focused discussion on the subject.

The Wrap Up

The lexicon may vary from patient centricity to observational research, contextual inquiry, or human-centered design. Nevertheless, the underlying ethos remains constant. We endeavor to understand people and their interactions within their day-to-day context. In essence, we seek to understand how we interact with the world around us.

Discovery research is what enables this understanding. We dig deep across a small sample to reveal not only behaviors and attitudes, but the "why's" behind them. We believe it's with this understanding that we can translate unmet needs into actionable opportunities—both products and services—to improve the patient's experience.

In the following chapter, we will look at the "ecosystems" in which patients live their lives, and how we understand them—and how they should influence the design of our products.

Chapter 2
The User's Ecosystem

Paul Blowers and Gavin S. Lew

Abstract In this chapter, we will bring the user's ecosystem to light. Designing within a health ecosystem involves not only consideration for the patient experience, but also situations and environments of use that contribute to a wider set of experiences, interactions, and relationships that, if designed well, can improve outcomes beyond safe and effective use. There are a myriad of touchpoints that make up the ecosystem in which a user will need to navigate in taking their therapy. These touchpoints vary and include their HCP, their pharmacist as well as caregivers. Case studies will be described to provide detail as to how some companies filled knowledge gaps and identified insight to advance therapies.

Keywords Ecosystem · Patient · User · Experience · Design · Environment · Journey

Introduction

Chapter 1 gave some framing on the research process—how to assess users' needs, and how to develop and assess concepts to address these needs. This discovery and development process is generally the same regardless of the kind of product or service, whether it be consumer electronics, public transportation, or a health-care product.

When designing products for industrial or consumer use, the one who directly interacts with the product is generally referred to as the "user"—whether they are interacting with a computer, ATM, kitchen appliance, or garden tool. In the case of a health-care product—as is the subject of this book—we often use the term

P. Blowers (✉)
Development Sciences, AbbVie, Inc., North Chicago, IL, USA

G. S. Lew (✉)
Bold Insight Inc., Downers Grove, IL, USA
e-mail: Gavin.Lew@boldinsight.com

R. R. Nesbitt et al. (eds.), *Patient-focused Development of Biopharmaceutical Products*, AAPS Advances in the Pharmaceutical Sciences Series 27,
https://doi.org/10.1007/978-3-031-98253-8_2

"patient" to mean the one who receives the drug product (by injection, swallowing, etc.). However, the reader may still see the use of the term "user" in later chapters, as there are others who interact with the patient (the ultimate recipient of the medication) to assist in their therapy. These may include caregivers—family or friends who assist in delivering or supporting the therapy. In either case, the focus of the book is no different—meaning the design should meet the needs of the individuals who will receive, handle, try to understand, hold, and deliver the drug products contained within. The key is—to take the designers' eyes off the internal (design, manufacturing) considerations alone and focus on how that product will be used (hopefully correctly) by its intended recipient.

This chapter looks specifically at users of a therapeutic to treat a disease, and better understand what is unique about the users' needs and the various perspectives that must be considered.

At the heart of any new therapy is the drug and its efficacy. And rightfully, the focus of new therapies should be toward efficacy. But, when efficacy is clinically demonstrated, integrating *human factors* into the product lifecycle is essential to improving health outcomes, because drug administration can be impacted by usability from the HCP (whether physician or nurse administering or training the patient or caregiver) to the pharmacist who plays a role in training and often connects to the patient, and the patient or caregiver experience. As drug delivery has evolved from pill-based therapies to more complex combination products and even digital diagnostics, emphasis on the *human factors* will be critical for success. Pharmaceutical and medical device companies have a greater need to understand and incorporate a much broader and deeper set of touch points. It is our belief that these elements form an ecosystem that not only requires understanding of interactions, but once these factors are understood, the system can be designed with better health outcomes. The commercial viewpoint described in Chap. 5 is reliant on a strong understanding of the ecosystem to also develop enabling messages and experiences for the customer.

In this chapter, we will bring the ecosystem to light. Designing within a health ecosystem involves not only consideration for the patient experience, but also situations and environments of use that contribute to a wider set of experiences, interactions, and relationships that, if designed well, can improve outcomes beyond *safe and effective use*. As shown in Fig. 2.1, there is a myriad of touchpoints that make up the ecosystem in which a user will need to navigate in taking their therapy. These touchpoints vary and include their HCP, their pharmacist as well as caregivers. Case studies will be described to provide detail as to how some companies filled knowledge gaps and identified insight to advance therapies.

To facilitate discussion, this chapter presents patient vignettes to facilitate a more humanized patient perspective that can link to practical design considerations to ultimately improve the patient, caregiver, and HCP experience. Suggestions for product and commercial teams will be discussed in an effort to demonstrate how to better design therapies to accommodate this ecosystem.

Fig. 2.1 The many interactions users must manage

Patient Vignettes

These vignettes illustrate points made in this chapter regarding the end-to-end experiences of a patient from initial symptoms through initial diagnosis and then through various treatment approaches leading to disease management. They expand on the overall life experience of the patients and how it is impacted by many facets of their ecosystems including family, career, personal interests, the medical system, and the fear, frustrations, and hope, as these come together to form an overall experience. Understanding the patient's journey and perspective as illustrated below is also key to establishing effective commercial approaches as described in Chap. 5.

Mary

First, we introduce you to Mary and her patient experience. Mary is 65 years old, married (her husband is a building contractor and an engineer), and has four adult children and several grandchildren. She retired early from teaching elementary school after 30 years of service. Symptoms have developed over time starting with twinges of pain in Mary's hands managed with over-the-counter pain management (analgesic creams and pills) as suggested by her family general practitioner. Otherwise, she is relatively healthy and maintains a busy schedule with volunteer work.

Seeking Answers

Over time, Mary's pain has increased. She finds it more difficult to type, write, and perform common household chores like washing dishes despite over-the-counter pain management. It has become the most noticeable first thing in the morning. Mary has become increasingly concerned and searches for information online to diagnose her problem. She's encouraged but also frustrated by the amount and diversity of information regarding hand pain. Arthritis seems to be a strong possibility. She discusses the progression with her family physician who refers her to a rheumatologist. Mary is relieved that she'll see a specialist and hopefully get some answers. Her anxiety level at the same time ticks up knowing that this is no longer something that may be easily managed and covered by her meager bridging health insurance until she reaches the mandatory Medicare qualification age.

Research Opportunity

Mapping the complexities of specialist referral and impact on patient/caregiver stress and patient health.

Patient adoption to a more advanced therapy involves decision-making factors that go beyond clinical diagnoses and often have underlying concerns, fears, and apprehension that can occur as the patient transitions from one medical practitioner to another specialist. Exploratory research that maps the patient's journey can identify areas of concern or confusion early in the decision-making process.

Method: Contextual interviews and possibly diary studies with small sample sizes. Focus questions on understanding the "in situ" experiences that are often identified while doing the task rather than having to rely on recalled retrospective memories. Having external artifacts present during data collection can also prompt more insight on the transitions that patients experience.

Stimuli: Main focus is on the patient journey and touchpoints.

Outcomes: Patient journey map with HCP touchpoints and decision-making moments. Understanding these interactions "in situ" or in the moment is vital to improving the connection. Capturing these moments while they occur tend to yield more accurate and rich data and is a useful experience sampling method [1].

The first appointment Mary can get with the rheumatologist is in 6 months. The doctor is extremely in demand and the clinic prioritizes Mary's initial consultation as less urgent. Mary, in the meantime, continues reading up on arthritis and writes a list of questions regarding what it could be and treatment options. She checks the health-care provider practice website, websites managed by medical institutions describing diseases and conditions, and company websites offering various therapies.

Research Opportunity

Medical information on the internet/social media and its impact on patient/family decision making.

Today, patients have choices, and often take the time to better understand proposed therapies. Resources are abundant online and getting the right information involves more than SEO (search engine optimization). Decision-making factors go beyond clinical diagnoses and often have underlying concerns, fears, and apprehension. Content provided to educate the patient needs to understand these factors in order to address these concerns. Exploratory research with patients who are candidates for more advanced therapies should be conducted. Patients who have successfully adopted can yield invaluable insights into the alleviation of patient fear and apprehension.

Method: Exploratory and contextual interviews with small sample sizes using one-on-one sessions.
Stimuli: Pharmaceutical websites, competitive therapy sites, and even consider starting with a web search.
Outcomes: Better understanding of what patients are looking to learn to craft navigation and content that is clear to support decision making and possibly mitigate fears with answers to patient questions.

In-Clinic Experience

The day of Mary's appointment finally arrives and she's excited to have an informed discussion with the doctor. She gives herself credit for not being a typical unprepared patient and has brought her list of questions. But the meeting with the doctor does not go to plan. She arrives on time but sits in the waiting room for 30 min with no explanation regarding the delay. Finally, she is called in by a nurse that escorts her to the exam room, collects vitals, and reviews symptoms. The nurse explains that the doctor is behind schedule due to back-to-back appointments and apologizes.

Research Opportunity
Physician office efficiency research.

Physicians, especially primary care physicians, often must see a high volume of patients a day to earn a living. While the experience in a clinic could be considered unique to the management within a given practice, the introduction of electronic health records (EHRs) or electronic medical records (EMRs) could create opportunities to evolve the patient experience. Consider the flags that adorn patient rooms to indicate patient waiting and status. As clinical settings move to EHRs, could the patient experience also change? Research inside clinical settings could uncover areas of improvement.

Method: In-clinic interviews where time and motion studies are conducted to capture HCP workflow and patient experiences. Observe clinician workflows and behaviors.
Stimuli: High external validity by using both in-clinic observations along with possibly data from the EHR on patient encounters.
Outcomes: Capturing bottlenecks and opportunities across daily patient encounters within a practice could identify changes in workflow to improve efficiency. The EMR itself could be more than a billing, ordering, and record system but setting the framework for a solution could make for efficiencies in the clinic.

Mary waits another 30 min in the exam room before the doctor finally arrives. The doctor gets right down to business with a perfunctory greeting and starts reviewing the electronic chart. She periodically asks Mary questions to confirm some details. Some of this is frustrating to Mary as she is being asked the same questions she already answered with the nurse.

The doctor starts an examination with little discussion regarding what she is seeing. She periodically asks additional questions regarding symptom progression. She abruptly ends the exam explaining blood tests are needed and she will get back to Mary in about a week. She rushes off and Mary has no opportunity to discuss her questions.

Research Opportunity

Aligning pharmaceutical material to support patient decision making.

Because physician encounters with patients are often time constrained, introducing new and often complex therapeutic options can take additional time. However, allocating time to ensure that the patient can understand the symptoms, diagnosis, and therapies is a priority. What can pharmaceutical companies do to help physicians be more effective in explaining possible treatment options to patients? The information must be aligned to the patient and research can guide that development.

Method: Using participants whose symptoms or disease progression make them candidates for a specific therapy, have not yet begun the therapy, but they are considering it as an option. Individual interviews that set the context and present material that will be reviewed by the patient.

Stimuli: Consider material designed for the patient, such as treatment material designed to help educate the patient and not the physician.

Outcomes: Understanding the mindset of patients at the decision stage of a particular treatment course can identify topics or areas where more information could be better presented to align to the concerns and feelings that patients have that cause a pause in decision making or simply help with questions to ask their physician. Content can be evaluated as accurate but still not address what would be useful in the moment. Properly designed material can help the physician, as well as the patient, with making critical therapeutic decision.

The Diagnosis and Initial Treatment Experience

Blood tests are performed, and Mary gets a call from the rheumatologist that the results are not definitive and recommends hand X-rays. This necessitates yet another trip back to the clinic. Mary hears back from the doctor a few days later with a diagnosis of rheumatoid arthritis (RA). The doctor explains it's an autoimmune disease where the body's natural immune system attacks the joints. Mary understands a lot of this given her research. The doctor prescribes a steroid and explains the side effects including weight gain.

Mary starts the steroid treatment and experiences some relief but soon she also experiences the weight gain side effect. This concerns her for both health and appearance reasons. Does she need to trade off hand pain relief with being overweight and its health impacts? She discusses this with the rheumatologist after a few weeks, and the doctor tapers Mary's medication. The pain increases, and there is little change in her weight. Is it time to look at an alternative? Will her health plan allow it?

Mary runs errands that include a pharmacy visit to refill prescriptions. She picks up several including her RA meds and expresses exasperation at the cost to the pharmacist behind the counter. The pharmacist points out that several of her medications qualify for rebates and manufacturer discounts through different programs. She pulls a pamphlet out describing these programs and discusses the drugs covered with Mary. Mary had no idea that all she had to do was ask about this but also wondered why the pharmacy didn't proactively offer these options when she got a new prescription. She signs up for all the qualifying discounts.

Research Opportunity

The impact of economic considerations on patient decision making, quality of care, and disease control/progression.

Costs associated with health care weigh heavily on patients and their families. Navigating the complex financials associated with new therapies can be difficult as payers often need preauthorization from insurance companies while some pharmaceutical companies provide medication cost options for qualified patients.

Method: Place calls to obtain preauthorization and cost information based on insurance coverage using a "mystery shopping" method of engaging in common insurance provider interactions.

Stimuli: Patient information that can vary to assess experience of obtaining preauthorization decisions.

Outcomes: Identification of patient information that is needed to make the preauthorization process more efficient and simpler.

For the pharmaceutical company who may have set up a special patient support center, the research would capture the needs of both the office admin in a clinical practice and the patient. Areas to streamline processing and reduce confusion can be identified with solutions for remedy.

Setbacks/Looking for Options

Mary becomes quite frustrated as the rheumatologist continues tapering the treatment with the continuing side effects and lack of improvement. She decides to get a second opinion and identifies another nearby rheumatologist recommended through a Facebook RA patient support group she recently joined. This, however, requires a discussion with Mary's insurance and approval to see a nonnetwork physician. Mary confirms that the visit requires an additional out-of-pocket fee.

Mary is pleasantly surprised at her initial consultation. The doctor is busy and a little behind but apologizes when he meets with her and starts the conversation by giving Mary adequate time to review her experience from initial symptom onset to recently failed treatment. He also sympathizes with the impact on her quality of life and relationships with friends and family. He reviews her chart, completes a physical examination, and begins to discuss alternative therapies. Mary is aware of these therapies through her own research and wants to discuss the side effects that appear to her to be very serious. She is also concerned that they involve injections, which is a significant escalation from the oral and topical treatments she is familiar with, and whether her insurance covers these drugs. Finally, she wonders aloud what if this doesn't work either?

The doctor answers her questions one by one clearly and succinctly and empathizes with her concerns. He reviews the details of the therapy including monitoring steps to detect and control side effects. He also explains that a pen is used rather than a syringe to simplify the injection process and that the injections are only required every few months. (Note: The term "pen" is often used to describe autoinjectors, which enable patients to inject the same drug product as in a syringe, but in an automated, much more easily operated fashion.) Her doctor does admit though that there's a chance this won't completely or may only temporarily relieve symptoms, but it has a great track record. This puts Mary at ease, and she agrees to try the new therapy. However, she will need to confirm it's covered by her insurance.

At this point, patients like Mary look to take advantage of the information available to them on the Internet. They search to learn more about the therapy. They go to the pharmaceutical website. They also go to aggregators of medical information, such as WebMD, Physician's Desk Reference (PDR), or specific clinical sites with strong brand reputation to provide reviews, such as Mayo Clinic and others. Or worse, they stumble upon less reputable blogs.

Research Opportunity

Evaluate pharmaceutical sites for clarity.

Patients look up information on their condition and possible therapies. While pharmaceutical companies cannot control where patients look for information, when patients do go to a pharmaceutical website, how can companies ensure the information presented is clear and addresses the needs of patients?

Method: Conduct formative usability testing. Bring in prospective patients for therapies and give them common tasks that the site was designed to answer. Consider contextual inquiry where before showing the website, patients describe how they deal with their condition and what concerns they have.

Also consider allowing participants to pursue information as they normally would—if this involves a Google search, then let them interact to learn about what they seek. Then redirect to the website of interest.

Stimuli: Pharmaceutical websites—consider both patient and HCP sites.

Outcomes: Formative usability testing will uncover how well the site answers the questions sought from *findability* (how easy it was to find the information) to comprehension. The goal is to identify content that was insufficient or confusing, as well as assess whether the navigation is clear such that patients do not wander into different areas and the confusion creates inaction over a therapeutic decision. From a business perspective: Does the site support the patient as it was designed to do?

Mary's new therapy requires one injection every 12 weeks using a pen. She is apprehensive about needles and injecting herself but likes the fact it's not another daily pill. Her doctor's nurse trains her on pen use for her first injection and Mary feels confident she can repeat the steps even though there are more than she anticipated. The nurse emphasized to be sure to hold the pen in place for several seconds after pressing the button to be sure the entire dose is injected. Mary thinks she understands but is concerned the next injection is in 3 months. Will she remember all of this?

Research Opportunity

Ensuring training material has the proper focus.

When auto-injection devices were introduced, many ways to train for proper device administration were created—from Quick Start Guides to training videos on DVDs. One study focused on assessing the training value of paper instructions with accompanying video or individually used.

Method: Qualitative research with individual sessions where different groups were given different training material (i.e., paper instructions only, video instructions only, and both). Participants were naïve to the drug delivery procedure.

Stimuli: Agency created material for use in the study and actual devices to assess successful administration.

Outcomes: Not surprisingly, the study found that participants who were given both paper and video instructional material performed better than only one material—either paper or video. What was interesting was how much video content was spent on the therapeutic benefits. Video content did not map to the patient journey of drug administration and instead added content suitable to the selection decision stage. Recognition of the patient stage is critical. Create content that is relevant to the stage of the patient journey (e.g., if the therapy has already been prescribed, content should focus on instruction how to perform a safe and effective administration rather than content applicable to the decision making). Research uncovered gaps in instructional material and was modified to reduce areas of uncertainty.

Patient's Real-World Medical Device Experience

It's getting close to the next scheduled injection and Mary picks up the medicine at her pharmacy. The pharmacist reminds her to put it in the refrigerator as soon as she gets home and asks if she has any questions. Mary had forgotten about this and had planned several other errands that warm afternoon. She had to scuttle those plans to get it home and in the fridge. Her injection is still a few days away, but she still feels confident she can do it on her own.

It's finally the day of the injection and Mary is nervous. This will be the first time doing the injection on her own, and her needle fear returns as well. She gets the drug from the refrigerator and attempts to open the carton to unpack the materials. The carton tab is marked OPEN HERE but is glued so tight that Mary cannot pull it loose especially with her RA flaring this morning. She resorts to a kitchen knife to pry open the box.

Research Opportunity

Ensuring a patient-centric carton design.

Another area to consider for patient centricity is when symptoms affect device interaction. An example is when a patient has physical or manual dexterity dysfunction due to disease progression that could affect interacting with a device or even the package itself.

Method: Formative, one-on-one research to allow for out-of-the box tasks where participants engage in opening therapeutic packaging. A first impression task prior to opening can assess package design and labeling, as well as reveal how one might open the package.

Stimuli: As near final as possible package design to allow participants to open and interact. Content on package could be in draft form if it does not interfere with opening. Be sure to include all materials inside the package, including the Instructions For Use (IFU), quick guides, and appropriately weighted device (in one example, the device was not available, but a weighted dowel was placed in a blister pack to simulate). This allows the first impression and opening to be more externally valid than when a patient interacts with an empty box.

Outcomes: First impression tasks uncovered how key words and images on the carton influenced how the patient thought the package was designed to be opened. Moreover, the organization of contents in the package were reorganized based on how the carton might be opened and when opened, how to prevent material from falling to the ground.

Once Mary opens the box, she glances at the instructions and gathers the recommended supplies. She remembers doing the injection into her thigh the last time. That went well so she will use the same location. She swabs her thigh with an alcohol wipe, retrieves the pen, removes the cap, and holds it down on her leg. She presses the button and feels the needle stick. Not too bad. She waits a few seconds, lifts the pen from her skin and liquid medication sprays against her leg. She then looks at the instructions more carefully and realizes that she lifted the pen too soon.

Research Opportunity
Auto-injector design for usability.

Using a device in your own home can be quite different from when a nurse showed how to use the device to administer the drug under the watchful eyes of a clinician. Steps that seemed easy in the doctor's office might not be as clear when one is alone. Moreover, those with experience using a similar but different device might miss steps because of their experience. These are all factors that need to be considered when designing a medical device and research can improve the design to improve successful drug administration when training is necessary and there is a time delay from training to first use.

Method: Training by an HCP is performed with patients in one setting and followed by a set time delay, which ideally reasonably simulates real life delay from training to first use (whether it is a day to a few days). As an example, participants are trained in one session and return a couple days later to administer the first dose. This can be done in a simulated environment, such as a research lab. Note that this method can be done in formative and summative validation studies.

Stimuli: All materials available to the patient in the real world should be used. Training should be based on written guidance with demonstration devices where applicable. For the subsequent test session following the time delay, participants will be allowed to interact with material expected in the real world—and this includes packaging, instructions for use, the device, and all other materials (such as alcohol pads as applicable).

Outcomes: For formative research, the goal is to identify areas of confusion and recommend modifications to the device, instructions, or packaging to improve successful administration. To improve the design, a highly tactical and targeted effort could be placed on different stimuli. For summative validation research, the goal is to no longer improve the design, but to identify and assess the risk of use errors that are observed when participants first use the device following a time delay.

Research Opportunity

Training decay.

Whether patient, caregiver, or even HCP, first time use can be challenging. The time between formal training and first use is an area for research. Mapping to realistic situations is ideal. For example, if the first dose of a therapy is administered in the clinic and the patient is expected to do the next dose the next day, having a 24-h separation from training and next use is appropriate. Other considerations would be shipping from a specialty pharmacy to the patient's home.

Method: Training sessions followed by a time interval of ideally 1–2 days can simulate a training decay.

Stimuli: Use production-grade stimuli and trainers who can simulate real-world scenarios.

Outcomes: Identification of potential areas of confusion associated with device, instructions-for-use, or training procedures.

Mary contacts the manufacturer support phone number printed on the carton after her experience. She explains what happened and that she noticed that medicine sprayed out after the injection. The call assistant walks her through the steps and asks Mary how long she held the pen before lifting it. Mary could not recall but thought she held it long enough. The call assistant tells Mary she should count to 10 before lifting the pen to be sure the injection is complete. The assistant pointed out this instruction for her and recommended she refer to it closely when Mary receives the replacement pen.

Research Opportunity

What can the adverse event reporting yield?

Reporting adverse events is standard from a pharmacovigilance perspective, but how can this data be used to improve design? In one example, a pharmaceutical medical device manufacturer wanted to better understand why a particular auto-injection device (or auto-injector, sometimes called a "pen") was reported to have "wet injections" where patients lifted the device before drug delivery was complete.

Method: Based on the limited information collected in the adverse event reporting, contextual interviews were conducted with 100 patients who reported a "wet injection" and asked for a replacement.

Stimuli: Past experience was the topic of the interview.

Outcomes: Over 15% of those who were interviewed said they normally administered the drug via a pre-filled syringe but were given an autoinjector instead. The pharmacist was out of pre-filled syringes and said this was the same dosage and briefly explained the injection procedure. Thus, the device type was switched at the pharmacy and patients were not trained adequately and an adverse event occurred. This insight led the pharmaceutical company to distribute more instructional material, including demonstration injection devices to pharmacies to better inform pharmacists that patients need the support for first time use.

Family Support

Outside of her RA, Mary has been healthy and has not required any health-related support from her family. She has, however, recruited her husband to help her, because she fears injecting herself. His background is in engineering, so she considered him more "technically-adept."

Research Opportunity
Test instructions for use (IFU) on patients and caregivers.

Therapeutic success sometimes depends on more than the patient with different user groups included in the research. Anticipated users and patients should also test material to ensure clarity and comprehension. Recognition that design is a process that requires iterative testing to evolve is critical because rarely is the first design sufficient.

Method: Formative usability testing on IFU with patients and caregivers. While some research involves "out of the box" studies where participants are observed using any material— whether all or none at all, the focus will be on the IFU, so participants will be explicitly asked to use the IFU to assist the injection process.

Stimuli: IFU and demonstration devices necessary for the administration of the drug.

Outcomes: Identification of use errors, close calls, and success with steps in the procedure, but also areas of possible confusion with content (both words and images or line art). These insights can be integrated into a redesigned IFU that is tested again and then used in a validation study with all product elements.

Claire

In addition to supporting her own health needs, Mary supports her widowed elderly mother, Claire, who still lives a relatively independent lifestyle in her own two-story townhouse about 30 min from Mary's home. Claire is 82 years old, also has RA, and suffers from weight-related diabetes and other conditions requiring a complex array of monthly, weekly, and daily medications. Mary helps her mother keep track of her medications and gets her back and forth to medical appointments when the senior shuttle is not available. In addition to her oral medications, Claire self-injects regularly using an insulin syringe for her diabetes and, with her family's occasional help, also uses a pen to inject RA medication.

Claire does her best to track her medications, but new ones are frequently added, and some medications have different loading doses or need to be weaned over time. Despite Claire's diligence, Mary is noticing problems.

Claire relies on pill reminder containers to sort her daily AM/PM oral doses. She uses a spreadsheet on her laptop but is still challenged to correctly sort the meds particularly when the dosing is irregular. For example, she has two meds that are once a week and one that is every other day. Mary attempts to monitor her mother's pill sorting but only gets around to it when she visits about once a week. Mary has noticed some errors and noted them. This frustrates and embarrasses Claire and puts strain on their relationship.

Research Opportunity

Calendar studies to test comprehension when dosing varies across time.

Patients often take much more than one medication and with the complexity of daily dosing, adherence risks are ever present. Dosing regimens sometimes vary. Consider a therapy where the initial dose is different than the weekly maintenance dose or when a drug requires a weaning or successively reduced dosage. How can the dosing regimen be clearer to patients and improve adherence? Some solutions involve blister packs with dosing orders or calendar-like features on the package itself.

Method: Formative research where participants are given a prescription and asked to perform tasks that assess comprehension. These one-on-one sessions will observe how participants use the material provided across a compressed calendar of time to assess dosing regimen. Limit context and information to only what is displayed on the package to ensure that only the information presented is evaluated. To simulate a time change, use of a clock to change the time or other environmental cues can artificially advance time within the research.

Stimuli: All materials that are expected to support the administration of a new therapy. This may include the IFU, quick start guides, as well as content and the design of the ordered blisters of pills on the package itself. Note that behaviors must occur rather than have participants verbally describe what they would do. Let them interact as necessary: for example, if a blister pack is involved, include placebo pills that participants will remove during the session. Ideally, have a fresh and new package for each participant. Let the behaviors emerge during the interaction rather than assume verbal assessments of what would happen.

Outcomes: Areas of confusion in the organization, design, and content of the instructions and packaging are identified.

One of Claire's medications requires mixing a powder formula with water to prepare a suspension for drinking once a day. According to the instructions on the label, the powder should be measured as a capful mixed with 8 ounces of water and thoroughly mixed into a suspension. Mary has noticed that her mother is using a kitchen tablespoon to measure out the powder and isn't very accurate or consistent with respect to adding the correct amount of liquid. Claire is also mixing it with orange juice rather than water to improve the taste and there are often clumped powder remnants at the bottom of the glass after she drinks the suspension.

With Mary's own new experiences with RA, she recognized that self-administering the RA medication might be difficult for her mom. With this in mind, she volunteered to help Claire with each self-administration. One unexpected problem early on was difficulty getting the pen to work in the seemingly softer—but preferred—injection location on Claire's abdomen. The pen has a mechanism to prevent accidental firing and requires using one hand to create a flat skin surface area. This is difficult to accomplish for Claire due to her size and, working with Mary, they have found acceptable alternative injection sites on Claire's legs.

Research Opportunity

Reconstitution studies.

Instructions for use should be considered an artifact to be tested because it is part of a drug delivery system. Consider processes where the patient needs to reconstitute (or mix) the drug prior to administration. The steps involved need to be evaluated for clarity.

Method: Test instructions for use early and often to ensure that the steps, images, and instructions are clear and to the point.

Stimuli: Instructions and functional devices to allow users to go through all steps in the process. Include illustrations and content.

Outcomes: Instructions for use improvements for comprehension based on observed user confusion. Possible modifications to the device and/or process to better align the reconstitution for accuracy and adherence.

In addition to her laptop, Claire has an iPad (a gift from her grandchildren) and recently stepped up to a smartphone after her flip phone broke and there was no comparable replacement. Claire is comfortable with the laptop since she's used a PC with Windows since the early 1990s when she worked as a bookkeeper at her husband's small business. Some of the new features on familiar applications like Word and Excel do confound her but so far, she has been able to use familiar controls to get around them although less efficiently. She's engaged in social media using Facebook to keep up with family and patient support group sites to help her cope with her maladies.

The iPad was a very different story. Claire relied heavily on her grandchildren to figure it out and understands the basic functions. She understands the concept of an app but only downloads games to entertain herself throughout the day. The grandchildren gave it to her for reading, but she's never adopted that and prefers printed books. The smartphone was easier to pick up, since it's like the iPad, but she only uses it for calls and texting.

Therapy Supportive Technology

Claire recently became aware of a smartphone app for her RA medication. Claire's nurse told her about it after she had complained it was hard for her to remember all the steps to inject the drug after several weeks and the instructions were too complicated. The nurse explained that the app includes a step-by-step review of the entire injection process so that she can watch before or while doing the injection. She also explained it has a feature to share her symptoms with the clinic. Claire likes the idea of the app but puts it off. She doesn't use apps on her phone and has never downloaded one before. After once again having trouble remembering her injection steps at her next dose, she breaks down and asks her daughter to help download the app for next time. Mary uses the app's instructional feature for her mother's next dose—and everything goes well. The app even reminded her to allow the pen to warm up for a more comfortable injection. She also likes the simple questions she can answer after the injection about her symptoms and is confident that they go directly to her clinic. The nurse contacts Claire shortly after the injection encouraging her to continue using the app for this so that they can more quickly respond with therapy adjustments or changes.

Research Opportunity

Software studies.

As software-based products from diagnostic to therapeutic support get into the hands of HCPs and patients, the need to ensure content clarity is essential. In a software app, however, there is more than just content; other features such as navigation, alerts, and other communications beyond the step-by-step instructions need to be considered carefully.

Method: Formative research where a moderator sets context for the participant about a prescribed medication or therapy. If this involves downloading new software to a mobile device or laptop, every step—from getting the app to opening it to using it to complete the drug administration—is to be evaluated in the study.

Stimuli: The software application and all accompanying materials that may inform and assist the patient to set up the software application. The app needs to have content ready, but also have navigation that is ready for testing as well. If reminders are part of the software, add these aspects of the software to the study Let the behaviors emerge during the interaction rather than assume verbal assessments of what would happen..

Software as a Medical Device

When developing products that include software in the health-care space, it is important to consider relevant regulatory guidance depending on the nature of the software, in addition to guidance on the application of human factors and usability engineering. For example, if the software in question has been previously marketed and is simply to be incorporated into a new medical device, the FDA has issued a guidance document titled Off-The-Shelf Software Use in Medical Devices [2] that should be considered because the software in question is itself is considered a medical device. There are various Software as a Medical Device (SaMD) guidance documents that should be referenced, noting that the way in which SaMD is regulated is evolving. From July 2017 to September 2022, the FDA conducted the Digital Health Software Pre-Certification Program. The intent of the program was to explore innovative approaches to regulatory oversight of medical device software developed by organizations that have demonstrated a robust culture of quality and organizational excellence and who are committed to monitoring real-world performance of their products once they reach the U.S. market. The primary conclusion of the program was that rapidly evolving technologies in the modern medical device landscape could benefit from a new (yet to be defined) regulatory paradigm, which would require a legislative change.

(NOTE: For more on Software as a Medical Device, see Chaps. 3 and 6.).

John

The next vignette concerns John, a 50-year-old airline mechanic with nearly 30 years on the job. Initially, John's work was hands-on, but over the course of his career, he has moved into supervisory roles and now spends most of his time managing a large team of mechanics at a nearby maintenance facility. That said, he still likes to get onto the shop floor regularly to work with his people to solve problems and share his expertise.

Early Symptoms

John has noticed some mobility issues while walking and while using his hands over the last few months. He is experiencing tremors that he never recognized before. More than anything, he is self-conscious that others will see his condition and question his stamina for the job. His wife also noticed the symptoms and asked him a few times about it. She is concerned that he's exhibiting symptoms for someone much older and recommends he see his doctor.

Hesitancy to See a Doctor

John rarely goes to the doctor, preferring to work through it when he's sick. He's been very healthy all his life and doesn't like admitting something could be wrong. One day, his symptoms became so bad that he just couldn't get himself out of bed to get to the office. He breaks down and schedules an appointment.

John's GP examines him and explains that his symptoms could be related to several possible diagnoses, but he recommends that he see a neurologist to narrow down the possibilities. John was hoping to get an easy answer and was alarmed he is being sent to a specialist. But he does not hesitate to schedule an appointment using the referral.

Seeing the Specialist and a Devastating Diagnosis

The neurologist's examination is more comprehensive and involves several physical ability and dexterity tests. John notes the difficulty he experiences completing some of them and his anxiety increases. The doctor explains that lab test results are needed for a final diagnosis, but she suspects Parkinson's Disease (PD). John is devastated. There is a history of PD in his family, and he has seen the debilitating effects of the disease firsthand. The doctor sees John's distress and explains that while there is no cure, there are many ways to manage it and delay the effects. The doctor prescribes oral medication to relieve some of John's symptoms.

Medication and Symptom Management

John notices immediate improvements once he starts his pill medication. His physical symptoms are reduced, and he can better manage at work and through his daily activities. There are some side effects of the medication, but he can live with these given the overall benefits. His neurologist advises him that the medication will likely become less effective as his PD advances and will require adjustments and that he may need alternative therapies later on. Over time, John's symptoms are becoming more pronounced despite adjustments to his medications. It's becoming increasingly burdensome to both John and his family to keep track of the prescription changes and maintain compliance over the course of each day. He needs to be careful; otherwise, he experiences prolonged periods of immobility and cognitive difficulty that make it impossible for him to do anything. This has impacted on his work as he can only come in when he is adequately medicated, and the symptoms are well controlled.

Wearable Therapy

John's neurologist recommends an advanced therapy that uses a wearable pump to deliver John's medication directly into the digestive tract. The doctor explains the primary benefits: (1) Drug delivery is more direct and effective than the current oral medications, and (2) the pump can be programmed to deliver the drug at an even cadence over time eliminating the pill dosing burden and potential for errors.

The doctor describes how the drug, in gel form, is delivered through a tube from the pump to a port surgically placed on John's stomach. The port provides direct access to the digestive tract. John will need to attach a drug cartridge to the pump once or possibly twice a day and John will need to keep the port clean.

Initial Concerns: Surgery and Stigma

John questions the risks regarding having a surgical procedure to establish the port needed to deliver this therapy. His doctor encourages him with safety information regarding this very common procedure but does concede there is some risk. John's other concern is the pump itself. He's sure he can manage the drug cartridge replacement and simple steps to control and monitor it, but he doesn't like the idea of wearing it all the time including at work and otherwise outside the home. He regards it as an immediate signal to everyone he meets that there is something wrong with him.

The neurologist has his nurse assistant walk John through the pump equipment and the steps required to run it, replace the drug cartridge, and special features to adjust the drug delivery. Afterward, John is more comfortable with the idea of this therapy and now sees it's a better alternative than oral drugs.

John completes the surgical procedure to insert the port and sets up the pump a few weeks later. He is trained on his pump before he heads home. He has the option to carry the pump using a strap or keep the pump in a discreet pouch he can wear under a garment. He prefers the latter but still does not like the bulk of it and how awkwardly it sticks out. It also makes noise that he knows others can hear making him self-conscious of the wearable device in public.

John notices considerable improvement in his symptoms with the new therapy and thinks he made the right decision. Still the equipment introduces new challenges and lifestyle conflicts.

Research Opportunity

Wearable therapy impacts on patient lifestyle.

Much of this chapter has focused on the therapy from package design to instructions to administration. But, in cases of therapies for chronic conditions, supplemental elements are meaningful to patients. Consider devices that are placed on the body but require attachments. What design considerations were

(continued)

made to allow a patient to go outside while wearing the device? Socialize? Work? Exercise? Understanding the needs of the patient, as well as, how well as the comfort for daily activities is critical to usage and ultimately, therapeutic success.

Method: Have participants put on a wearable product. Assess first use for initial fitting, comfort, and ability to perform daily functions related to drug delivery, but also mobility independent of therapy. Because therapy may be chronic, consider longitudinal assessments where the participant wears the device across several days or a week to understand longer-term impacts of different environments from sitting in a vehicle to walking outside.

Stimuli: Wearable device that is fitted for the patient with a functional device.

Outcomes: Initial fitting and removal / reattachment to understand comfort and mobility. Use of drug delivery or diagnostic (e.g., can administration easily be performed / can information or are indicators visible while wearing the device?) Longer-term usage may reveal product design insight beyond initial use.

Bonus

Adhesive Use: Consider longitudinal wearing of device that requires an adhesive to secure where assessments are performed after initial application but also 24 hours or several days later if the device can be attached for this length of time.

John has had to rethink his routine around the house now that he wears the pump most of his waking hours. It's important to immediately restart the therapy first thing after getting out of bed to start managing the symptoms. But this is often the worst time of day for John's mobility. His tremors can make it very difficult to attach the pump to his gastrointestinal (GI) tube, put it on, and start the pump. His wife, Christine, often needs to do this for him, which makes him feel dependent.

John is mostly able to manage on his own around the house once the medication kicks in. He loves to work on his vintage car in the shop and needs to be careful about machinery and tools around the pump. He wears the pump and keeps the tubing under his clothing to be sure nothing gets caught or pulled.

John recently noticed redness and pain around the GI port. He called the clinic and was asked to come in as soon as possible. The area around the port had become infected, and he was reminded how important it is to keep the site clean. This made him feel like a child being scolded.

Research Opportunity

Trust but verify (a story of applesauce and gravity).

(continued)

Research on the patient and caregiver experience is critical to understand the experience, but it also can verify the engineering specifications. Consider a study where medicine is delivered from a container via tube to the patient.

Method: Formative, one-on-one research to validate drug administration based on product design specifications. Participants are given instructions and tasks to perform the administration procedure.
Stimuli: Actual material used in the process from mannequins with intake ports to tubing to fluid to administer (with viscosity similar applicable drug, and in this case, applesauce).
Outcomes: Surprisingly, in one case example, dry run preparatory research activities found that the fluid (applesauce) would not flow to completion. Checking with the product engineers, who were perplexed with why issues were found, the actual engineering report procedure was reviewed. Gravity was the main force that moved the fluid to the body, but for the report, the procedure aligned the tubing straight up and down. But in real life, the tube is coiled in the package and certainly not straight but twisted. Daily activities involve moving around, which can change the direction of the tubing (such as laying down) and/or irritate the port site. In this example, what worked in the engineering lab failed in research.

Other Influencing Factors: Mental Health

Christine has supported John from the moment of his diagnosis. She accompanies him to most medical appointments and keeps notes regarding examination and test results and modifications to his treatment regimen. Christine sees the physical, mental, and emotional toll PD is taking on her husband. She is often distraught but tries to maintain a brave face. She keeps very busy helping John with his daily activities, collecting his prescriptions, helping him fill his pill organizer, and monitoring him for any changes she thinks are important to note for the doctor. With the addition of the pump therapy, she helps John first thing in the morning get it set up and running. John is usually at his worst in the morning and could not function without her. He is incredibly grateful but hates being such a burden to her.

Other Influencing Factors: Pets

John loves animals and has always had at least one dog in the house. He adopted a puppy German Shepherd mix shortly before the PD diagnosis and looked forward to training her and using the dog as an excuse to get more exercise. Although his symptoms are better managed now, he still finds it difficult to give the dog attention including regular walks. The large dog seems to get into everything and likes to chew; so, John must be extra careful about keeping his medications and pump equipment well out of the dog's way.

Other Influencing Factors: Change in Environment

Recently, John had to figure out how to keep his drug cartridge supply refrigerated during a trip to visit his grandchildren in another state. He had to plan and arrange to pick up drug cartridges locally. He also had to be sure that the hotel offered in-room refrigerators.

Global Considerations

Much of this chapter's topics can be expanded to global markets. However, there may be greater differences in patient perceptions and fears, as well as completely different influences on individual health-care ecosystems. Consider this vignette of Amara, a woman in a rural African village who becomes ill and seeks medical attention. Unlike the pattern used previously in this chapter, research was already underway to understand why individuals request multiple "first-time" HIV tests, and this important finding was uncovered during the study.

Global Research Finding

To understand how medical diagnoses are made in one country in Africa to uncover why individuals seek "first time" HIV screening tests multiple times, researchers conducted a study in the region of interest where HIV is prevalent, and free HIV tests and medication are available. However, excessive amounts of HIV tests were being given: more than the population of the communities in the region of interest, which sparked the need for further research.

Method: Researchers accompanied participants on a health-care journey. Contextual interviews and direct observation occurred with a small qualitative sample.

Stimuli: Observations of interactions within a particular ecosystem.

Outcomes: Discovery of social stigma surrounding HIV positivity and fear to the point that it disrupts patient care and adoption of therapy.

In the study, researchers found a poignant experience with Amara, who went to a local clinic where the physician tested her for HIV based on her symptoms. Unfortunately, she tested positive. The physician discussed advances in treatment options and that they were available for her. She was given a green folder with her information and told to simply wait in line for free medication. She went outside and noticed that everyone standing in the queue was also holding green folders. She then looked across the street and saw the local market that she frequents. She leaves without medicine, because she fears being seen by someone she knows in her community. Later, her symptoms worsen, and she walks many miles to another clinic in a different village and another HIV test is performed so that she can get treatment.

Previously unidentified societal stigmas, uncovered through this research, can broadly influence patient care and disrupt ecosystems, potentially negatively impacting population health and outcomes. Going beyond the test or the medication to understand how it fits into the patient ecosystem is essential. Designing around the patient's experience and ecosystem is key to ensuring adoption and adherence.

Conclusion

The intention of this chapter was to explore the multitude of decisions and touch points that go beyond therapeutic efficacy. Considering these factors when developing new therapeutics is essential. Success in today's medical landscape needs to account for a broad ecosystem—one that considers more than clinical efficacy, but the patient experience, which extends to caregivers and HCPs, environments, and situations of use. Designing the next generation of therapeutics is about more than choice or information or even usability and human factors associated with the drug administration and delivery. It is about designing a therapeutic experience that will have high rates of compliance, because its development incorporated a litany of factors relevant to the patient. Chapter 5 expands on this discussion, illustrating commercial approaches to both understand and draw important learnings from a strong understanding of the patient ecosystem.

Using the vignette as a vehicle to trigger a patient-centered approach to design adds dimensions that are often overlooked or assumed. These dimensions include (but are not limited to) the following:

- *Dimension 1*: Pre-existing experience—What life was like before the disease. This is often a patient's and the patient's loved one's aspirations after the disease strikes. There are freedoms taken for granted like general mobility, participating in hobbies, travel they want back. How can the therapy and its regimen be designed to partially, if not completely, restore life before the disease?
- *Dimension 2*: Acceptance of current situation. Recognition of how early symptoms can be relegated, such as patients may ignore or overlook early symptoms blaming them on age, for example. Getting help may be admitting loss of control or could be delayed for fear of the unknown. Early medical consultations can be confusing and frustrating as physicians diagnose and pinpoint a cause. What information can be useful at this stage of the decision-making process to reduce confusion and fear. What questions are applicable when they are sent to a specialist that may mitigate overall patient anxiety?
- *Dimension 3*: Diagnosis and initial treatment. This may be a time of both relief and anxiety for the patient. There is an identified medical reason for the pain and discomfort but what will that mean for their future?. Therapies discussed at this

stage come with hope but potentially side effects and only temporary relief for progressive diseases. Patients may be confused by the prognosis and more complex therapies. They seek support from family, the HCP, and, increasingly, social media.

- *Dimension 4*: Medication management and the pharmacy. The patient's ecosystem may include multiple medication therapies covering comorbidities beyond this one most recently diagnosed. Patients and caregivers have been observed in research with dozens of prescription bottles. In some cases, these are neatly and diligently managed using calendars and pill boxes, while in others, patient's struggle to remain compliant with the barrage of drugs. The pharmacy is often a lifeline not just filling prescriptions but suggesting cost-saving payment options and other alternatives.
- *Dimension 5*: Setbacks and looking for options. Patients may experience setbacks along the way requiring multiple dose changes and a series of medications as the HCP works to fine-tune or move onto other more effective therapies. This can be a confusing series of changes for the patient with often little apparent benefit. The patient may search themselves for alternatives through social media, patient support groups, or just "roam" the internet. This can be a challenge for the medical management team to balance patient expectations, therapy risk/benefit, and cost.
- *Dimension 6*: Technology enablers and challenges. Patients often have access to technology that may be leveraged for increased compliance and active involvement in the therapy and their wellness. Many patients use, for example, social media and browser capabilities to reach out to others with similar medical issues and search for disease and therapy information. However, other patients may be technically naïve and either use it minimally or not at all. Any attempt to leverage technology in the patient ecosystem must account for this segment to provide alternatives and hand-holding to overcome patient hesitation.
- *Dimension 7*: Learning to use a medical device. Patients are anxious when first learning a medical device, particularly those that are even minimally invasive (e.g., include needles). As mentioned previously, many are not technically adept, so simplicity is important to learn and remember how to use the device. The best learning stems from well-designed training and instructions that consider the patient's ecosystem (as illustrated throughout this chapter), perspective of the device, health literacy, education level, and cognitive/physical capabilities, and includes hands-on demonstration and practice. Repeated training and instructional reminder tips are often necessary for more complicated devices or devices that are used infrequently as patients may forget steps.
- *Dimension 8*: Wearable therapy. Patients with a wearable device are concerned about using it correctly, the impact on daily activities, and privacy. Wearable devices are more complicated than standard oral, or syringe-injected medication potentially requiring programing, connecting to infusion lines, periodic adjustment, medication loading, and sterile procedure for ports. Training needs are high, and device design and instructions must address real-world complications introduced in the patient ecosystem (e.g., sleeping with the device, non-sterile

environments). When worn, patients want a device to work for and with them as they go about activities of life. This includes dressing, completing chores, hobbies, exercise, interacting with family and pets, and travel. Most patients want a device to be invisible to the outside world that otherwise may serve as a visual sign that something is wrong with them. Size and discreet location of the device must be considered in the design to meet these concerns.

- *Dimension 9*: Real-world challenges to the patient adhering to therapy. Patients face both significant and mundane obstacles when managing the medication regimen. They often have multiple health issues requiring various medications with potentially conflicting dosing regimens. Many organize medication regimens using pill reminders, paper diaries, and digital tools like spreadsheets. Complexities are compounded by patient disease and age-related cognitive difficulties leading to dosing errors and the impacts of noncompliance. The extra burden to ensure compliance is often placed on a patient's home support system (see next dimension).
- *Dimension 10*: Family support. The patient's spouse, children, and other family members are often closely involved in the care regimen for their loved one. It can be stressful particularly when the caregiver(s) continue to have other commitments including jobs that leave them little time to rest. It may take an emotional toll seeing a loved one suffer with a disease and deal with individual anxiety regarding the outcome. The family caregiver takes on many of the medication management and administration tasks for the patient and can therefore be subject to various omissions or other errors given the stresses described. They must also be considered primary users in addition to the patient when designing highly usable medication administration devices and regimens.

Take these dimensions under consideration to help describe that complex ecosystem of many moving parts that contribute to patient care and health outcomes. How can therapeutic design incorporate these dimensions to improve patient outcomes? Having an effective therapy for treatment isn't sufficient. Therapeutic success depends on maximizing proper administration and adherence. Understanding these patient dimensions and designing for the patient is key to success.

Figure 2.2 shows the array of factors that impinge upon a user (the patient and a caregiver). The reality of the complexity of care creates an ecosystem of interactions and relationships. Given that complexity, what can be done to influence therapeutic success? How can we, as practitioners, effect a positive change to the elements of the ecosystem?

As this chapter has clearly demonstrated, we believe that ***good design matters***. We capture insights to better understand how HCPs, caregivers, and patients interact and then design a solution that will fit those needs. This allows the product and commercial teams to actually *shape a more intuitive process of how drug administration can* fit the ecosystem to promote safe and effective use, bolster adherence, and anticipate patient needs.

In the next chapter, we will look at the users' needs in drug delivery, and how to better understand and address them.

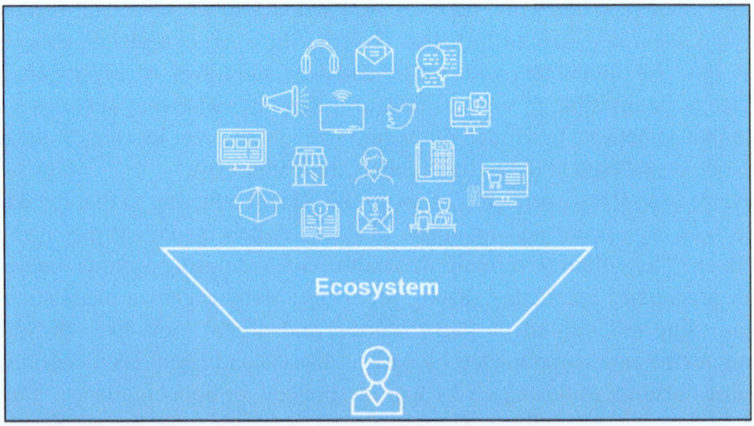

Fig. 2.2 A user must manage all the other stakeholders as well as all the components in their therapy

References

1. Lew, G.S. (2009). What do users really do? Experience sampling in the 21st century. In: Jacko, J.A. (eds) Human-Computer Interaction. New Trends. HCI 2009. Lecture Notes in Computer Science, vol 5610. 314–319. Springer, Berlin, Heidelberg.
2. Guidance for Industry and Food and Drug Administration Staff" (August 2023, Docket number FDA-2019-D-3598)

Chapter 3
User Needs in Drug Delivery: Unique Challenges with Combination Products

John Towns, Andrew Susott, and Muriel Didier

Abstract There are many unique considerations and challenges that patients and caregivers face in appropriately delivering an accurate dose of medication when packaged as drug/device combination products. Careful thought must be given to the early complementary product development plans of both drug and device when these are combined together into a single marketed product. Special attention should be given to patient-centered design aspects and to the execution of use studies that properly evaluate the combination product in simulated and/or real-life use environments/situations. Consideration of the user needs should be incorporated promptly into the design and development of combination product, along with the focused development and testing of clearly and easily followed instruction for use. An emphasis on user needs in the early development and subsequent use studies of the whole product (combination product and instructions) provides greater assurance in the understanding and proper use of the combination product to effectively deliver medications by patients and caregivers.

Keywords User needs · Combination products · Patient-focused design · Connected care · Pediatric considerations

J. Towns (✉)
Regulatory Affairs, Eli Lilly and Company, Indianapolis, IN, USA
e-mail: mjtowns1@comcast.net

A. Susott (✉)
Global Device & Packaging Development, Novartis, Durham, NC, USA
e-mail: andrewsusott@gmail.com

M. Didier (✉)
Novartis, Basel, Switzerland
e-mail: mail@murieldidier.com

R. R. Nesbitt et al. (eds.), *Patient-focused Development of Biopharmaceutical Products*, AAPS Advances in the Pharmaceutical Sciences Series 27,
https://doi.org/10.1007/978-3-031-98253-8_3

Introduction

In the development of drug/device combination products, the patient's ability to successfully receive their medication by way of a well-understood, patient-focused user interface should always be at the center of a sponsor's attention. As the drug product is likely self-administered (orally, injection, or topical administration), the patient becomes the sole determinant in the (1) successful self-administration of the medication and (2) the continued adherence to the ongoing drug therapy dosing schedule. Note that we use the term "adherence" to be comprised of two parts: persistence (the continued and prolonged habit or existence of something) and compliance (the degree to which a patient understands and correctly follows medical advice).

Patient-focused, or patient-centered, design of products is generally accepted as ensuring that patients' experiences, perceptions, needs, and priorities are meaningfully incorporated into development plans and decisions relating to their health and management of a disease. In this chapter, we focus mainly on injectable combination products, as many medications require the route of administration to not be simply taken orally (the most convenient, straightforward delivery of those medications), but through a medical device constituent part (the portion of a combination product that involves the medical device or delivery system). The drug/device user interface, which includes secondary packaging and instructions for use, forms the patient experience as a combination product for use by the patient to best administer the drug dose. How the patient, caregiver, or health-care professional interacts with the combination product is of critical importance in assuring a complete dose is received [1]. Although the main emphasis should be efficacy of the drug in treating the disease state, the patient's handling of the combination product has gained attention [2]. The focus is to ensure the best patient experiences in use of the product, including the injection is received within a reasonable timeframe (5–15 s). The primary mode of action is the drug in the treatment; however, the device constituent part that delivers that drug is what the patient holds and interacts, with device drug delivery design outputs being key to ensure delivery of the intended drug dose to the intended delivery site [3]. The patient will remember how convenient, or not, the medication dosing experience was—this may impact adherence of that patient remaining on their medication schedule—for good or bad. This is all in achieving the goal of helping patients in managing their medication regimen resulting in favorable outcomes in the treatment of their disease.

The Patient's Use Needs

The focus of this chapter is twofold. The first focus is the recognition and appreciation of patient capabilities to successfully self-administer their therapy via combination products. The second focus is the advantages and challenges of timing the introduction of the to-be-marketed combination product into the early phases of development. This chapter touches upon the drug development timelines and explores the patient-focused considerations in the evaluation of the drug/device combination products in the context of the drug constituent part taking the lead as providing the primary mode of action in treatment of disease, with the device constituent part's role in successfully delivering that medication to the patient [4–7].

There are several attributes of the combination product user interface that play a significant role in optimizing the patient experience in the successful administration of a complete dose of medication. With the device constituent part being the primary patient user interface, optimization of patient-focused design elements for the user interface assures a satisfying patient experience in the administration of the drug dose. Examples of more progressive design considerations in the advancement of the most desirable patient-focused combination product user interface include: a convenient (e.g., infrequent) dosing schedule, high device tolerability (e.g., low pain upon injection), ease of use/preparation (e.g., few steps), ease of drug delivery (e.g., single injection, on-body wearables), feedback mechanisms of dose completion (e.g., audible click, visual plunger placement, time clock), and low injection anxiety/exposure (e.g., hidden needle).

A further emerging trend in patient-focused design is the move toward connectivity between device sensors and mobile device apps to the current drug/device combination product. These additions create a wide array of ways in which patients can better manage their medication schedules, keep track of doses taken, and may increase the likelihood of adherence in completing the full course of therapy. The collection that the patient requires to get the best therapy may include multiple devices, sensors, and transmitters embedded into existing combination products. Mobile apps are used to display important, real-time, personal, health-related information on watches, phones, tablets, and laptops.

Drug Development Timelines

Modern drug development requires a very rigorous and thorough development path, which can take from 8 to 12 years. Although this simplifies the work, the following are the major stages of activity:

Early Drug Discovery In this first stage, basic research on molecule's actions leads to potential targets, which is identifying biochemical sites of action in the body to which this candidate molecule might have an effect. If such an effect is determined, it is called a "hit" (i.e., the molecule does impact the target). Next, these hits are narrowed to a leading candidate (or "lead"), which seems to provide the highest likelihood of impacting the target and effecting a change. These candidates are then taken into the following steps:

Preclinical Studies In this next stage, the potential candidate is taken through more extensive in vitro (literally, "in glass", i.e., lab bench) testing to determine the action of a molecule. In this phase, there are studies of its Absorption, Dissolution, Metabolism, and Excretion (ADME), to better understand how the body might process the molecule. As better understanding of the potential benefit of a molecule is realized, the next steps are to conduct in vivo (literally, "in a living being") testing, in which the drug candidate is tested on animals, called preclinical research. In this stage, there is further work on the formulation, with the intention of improving its bioavailability, or a measure of the molecule's ability to impact the body and effect a change. During this phase of work, dose-ranging studies are conducted, which enable determination of the appropriate quantity of drug to administer. This will improve the molecule's ability to be accessible to the body in an effective manner. When it can be determined that the molecule is safe and effective, and an understandable impact on a disease can be measured, an application to submit the candidate for use in humans is made. In the United States, by way of example of a regulatory region, this is through an Investigational New Drug (or IND) submission to the FDA (or Clinical Trial Application for EMA).

Clinical Development Upon approval of this investigational application by a regulatory body (such as the U.S. FDA), the candidate is moved into clinical testing (i.e., testing on humans):

Phase 1 The first human testing is called Phase 1, where the candidate drug is tested on healthy volunteers, to ensure safety and a safe dosage range in humans. The drug is tested in a small group of 20–100 people with studies often run as dose escalation studies. The initial dosing starts out slowly and would likely be subtherapeutic, but with ascending doses being administered over time. Phase 1 trials aim to determine the best dose of a drug candidate with the fewest side effects. If review of this data does not indicate safety concerns, then the drug is able to move to the next phase of clinical studies.

Phase 2 In this phase, patients with the disease indication, which the molecule is intended to treat, are enlisted to be subjected to studies of the efficacy of the candidate. There are typically multiple Phase 2 studies, aimed at gathering data on different stages of disease progression, and different doses or dosing (these can be called Single Ascending Dose (SAD) or Multiple Ascending Dose (MAD)) studies, which are aimed at better understanding how the drug should be dosed—not just "how much" (quantity, in micrograms or milligrams) but "on what schedule" (should there be a range of doses across time). In this phase, ongoing pharmacokinetic assessments are made (i.e., characterizing the timing and extent of the drug on the patient), as well as studies to confirm the best delivery method and how to measure the impact on the body. These bioanalytical methods are developed and validated. At the end of the Phase 2 studies, the assessment of all the above data, and successful achievement of the primary clinical endpoints, the sponsor is confident that the drug product is safe and effective in humans with the intended disease indication. The sponsor has a clinically substantiated dose and dosing schedule by which the drug should be administered and has outlined a plan for more extensive clinical testing on patients with the disease state. Often during Phase 2 studies, usability testing is conducted (via formative Human Factors studies) to better understand the users' needs and to test different product presentations. At this "End-of-Phase 2″ (EOP2) point, the drug candidate is moved to Phase 3.

Phase 3 Phase 3 studies expand the administration of the drug to more groups and a larger number of patients with the indication for which the molecule is intended. These studies often include multiple "arms," each of which have a particular design, sometimes different dosing regimens, sometimes different groups or patient demographics. Phase 3 continues the pharmacokinetic assessments and expanded bioanalytical methods development and validation. Most regulatory bodies expect that the delivery method intended for ultimate commercial presentation post approval will be studied and data collected using this "to-be-marketed" combination product presentation. For example, if the Phase 2 clinical data was conducted with the drug administered via a prefilled syringe, and the firm intends, upon regulatory approval, to launch the product to the market in an auto-injector, regulators generally expect the phase 3 study would include a demonstration of bioequivalence, or demonstration of consistent clinical performance in the two delivery formats. At the same time that clinical data is being collected, the design, manufacturing, assembly equipment, and processes, which will be required for the delivery system intended for the commercial launch is being developed and validated. In addition, while Phase 3 studies are conducted, usability testing is typically done with users of the product in its final form (i.e., including all components, any labeling, or instructions for use) to ensure the users of the product can operate the system safely and deliver the dose as expected. This usability testing work (referred to as "Summative" studies or human factors validation studies) is a part of what is submitted to regulatory bodies to confirm that the users can safely and effectively utilize the product in its final, commercial product presentations. When the phase 3 clinical studies have been able to achieve their targeted endpoint, and other manufacturing, development, process, and usability validation work is completed, the firm will prepare their submission to the regulatory agencies that oversee the intended regions. In the United States, the submissions to the FDA may take the form of a New Drug Application (NDA) for small molecules, an Amended NDA (or aNDA) for a line extension of a previously approved small molecule, monoclonal antibodies (mAb), or antibody drug conjugates (ADCs). Upon approval by the regulatory authority, only then can the drug product or device be launched (in those regions where regulatory approval has occurred).

Phase 4: Although it is not as widely discussed, there is a Phase 4 to which regulatory bodies may often refer. In that phase (i.e., post launch, once marketed), there is an obligation for the firm to collect product complaints related to the device and adverse events (AEs) related to the drug and be prepared to manage the product to mitigate these risks. There may be additional clinical studies the firm conducts to seek approval for additional disease indications not covered in the clinical data supplied in the initial submission and approval.

Drug and Device Development Paths

In the study of the patient-focused user needs, there is a tension between the two development plans—drug and device—prior to the two constituent parts coming together to form the combination product. Up to the point of assembly together, the two parts are often developed separately in distinct disparate development paths. At a certain point in the clinical plan, the device is introduced, and these two constituent parts are combined into the continuing development program. An analogous description of the development path of a combination product is as a long major river that builds as it flows downstream and out into the ocean (see Fig. 3.1). This river is the development plan for the drug constituent part where the flow is the phases of development—a relentless, unwavering outpouring that moves at a relatively constant rate downstream toward, in this tale, product submission, and approval. The device constituent part is like a boat launched onto this steady flowing drug river. And like a boat on a body of water, the device constituent part progress is largely dictated by the direction and rate of speed of the drug constituent part. How, where, and when the device boat is put out onto the river will determine how the drug and device constituent parts are used together as one, evaluated properly during development and eventually included as the to-be-marketed presentation(s).

The device boat analogy of bobbing along on the rapid drug river current serves to illustrate a major challenge in the development of the combination product. That is, the drug development is often up and running at full speed before the device is introduced into the clinical trial (CT). This can be as late as Phase 3, where the clinical trial protocols are focused to address the drug claims. The device constituent part is often introduced late into the combination product development plan with little time for evaluation of patient contact and ability for revision to optimize user needs. This points to the importance of early-stage device technology development. This can often be a distinct path independent of drug but is of great value in mitigating the risk and impact to the overall development timeline. By the time the optimum drug-device attributes are identified, the device is at the proper level of advancement to progress for more patient-centered design and overall product development optimization. Two challenges with the merging of the drug and device

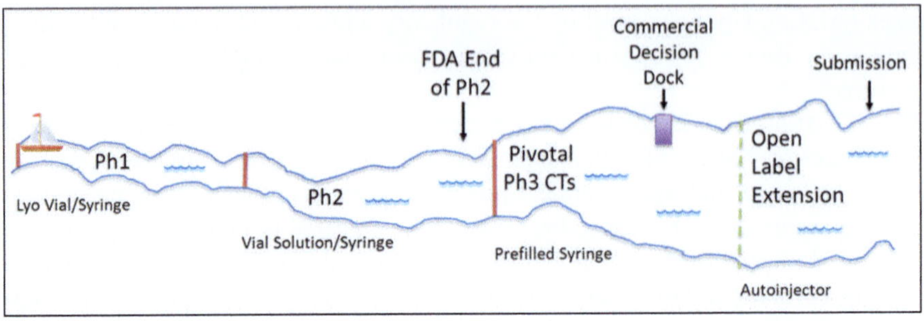

Fig. 3.1 Illustrative launching of the to-be-marketed device constituent part onto the drug river

constituent parts into the single combination product are illustrated in Fig. 3.1. The first challenge is the timing of the regulator/sponsor interaction (e.g., FDA End of Phase 2 (or EOP2) meeting) when seeking agreements of the Phase 3 clinical plans in supporting the eventual approved combination product label claims. At the end of Phase 2 stage in the drug development plans, the Phase 2 safety and efficacy progression are established for the drug in the treatment of the disease. However, the device constituent part to-be-marketed has not yet made its way into the clinical development program. Thus, the clinical drug/device presentation may not be representative of the eventual to-be-marketed presentation but may be the most expedient and cost-effective for the early clinical evaluation of the medication as it progresses through the development prior to making commercial decisions. The to-be-marketed presentation(s), in this example the prefilled syringe (PFS) and/or autoinjector, is (are) yet to be introduced, with the drug in Phase 2 withdrawn from a vial and delivered by an "off-the-shelf" syringe—likely by a health-care professional. There has been no experience gathered yet on patient self-administration for the eventual home use of the to-be-marketed product. This makes it difficult to seek agreement with regulators on the acceptability of the to-be-marketed products without any supporting data. In the USA, this often just leads to FDA responding at the EOP2 meeting on device constituent part issues as: "This will have to be a submission review issue."

The second challenge is the timing (placement) of the Commercial Decision team's availability to aid in gaining patient-focused experience for the to-be-marketed presentation(s). The commercial team formation is often quite late in the overall development path as the process from development to commercial launch is based largely on the drug timing, not on the overall combination product. By Ph3, the clinical trials have been hashed out and will meet the desired drug label claims (the river is naturally flowing toward the sea); however, the device constituent part may just be getting started. There is often little opportunity to generate patient-focused study data in support of optimized user interface and overall patient experience and satisfaction. As a result, the developer must rely on simulated use human factors studies as the best way to capture the patient impact of the system being developed. As shown in this illustration, the Commercial team is often assembled way downstream near where the pivotal Phase 3 clinical trials are almost completed. This timing is fine for the drug constituent part, as the clinical data in support of product label claims for safety and efficacy in the desired populations is on course for delivery into the submission. However, the timing of assembling the commercial team is just too late for the device constituent part and often has the device development teams "scrambling" to catch up to enact patient-focused commercial study design and execution in support of the submission label claims and user interface for the to-be-marketed presentation.

This timing issue frequently arises for the device constituent part as the development plan nears its end in Phase 3 and approaches initial submission. Too often, by the time the pharma firm's Marketing organization gets involved, the user interface has been designed, developed, and finalized and any request for changes the Marketing group requests may conflict with use error controls, mitigations, and—if pursued—endanger a successful Human Factors validation. Many device development evaluations and decisions still need to be accomplished with human factors summative studies in the validation of device stability/reliability and real-life usability by patients in home use, self-administration of medication, optimized instructions for use, use of quick reference guides, differentiation of like-products, and secondary packaging (e.g., cartons).

This all points to the need to introduce early into the clinical trial program the device constituent part of the combination product, including the entire product presentation (i.e., the drug delivery device, instructions for use, quick tip guide, and carton—including reference codes and apps). Although development teams will likely want to defer the development costs of device manufacturing and resupply costs as long as possible, the benefits of gaining patient-focused experience for these design attributes will likely, in the long run, outweigh the logistical costs of delaying introduction of the to-be-marketed product into the overall development plan. The challenges associated with merging the drug and device development plans explain why many sponsors (pharmaceutical firms) find it advantageous to have the initial launch of the medication in a first-generation device constituent part (e.g., prefilled syringe) and then follow up with a subsequent product launch with the second delivery system (e.g., autoinjector).

Considerations in Patient-Focused Design

Dosing

Selection

Given that not all aspects identified as important to patients, caregivers, and clinicians can be addressed by an investigational treatment or be measured within the clinical trial program, careful thought should be given to dose selection. As described in the above section, it is important to introduce and evaluate these patient-focused product designs as early as possible into the development plans and clinical trial program. This section addresses the issues related to selecting what to measure in a development program and identification or development of fit-for-purpose clinical outcome assessment (COA) to assess outcomes of importance focus to patients/caregivers.

Administering a Complete Dose

A major concern of sponsors and regulators alike is ensuring the complete dose of medication is delivered to the patient. A major focus is to develop the combination product that is so intuitive in design that it can be easily handled throughout the drug delivery process, minimizes patient discomfort in receiving the dose, and is not overly burdensome to complete the dose (e.g., excessive delivery time). The intuitive design may have to overcome the preconceived concept of how to operate the system based on the experience the user had with a previous system (called "negative transfer"). The design should also take into consideration clear feedback from the system that the complete dose has been delivered. This can be accomplished by a visual, sensory, and audible cues. In addition, often the instructions for use instruct the user to keep the device on the injection site for a set time following device activation to ensure the complete dose is delivered.

The assessment of the length of time required to deliver a full dose is an important one. This can be a concern on both sides of the injection time spectrum—either too fast or too slow injection times. Too rapid an injection can contribute to pain upon injection, whereas too slow an injection can adversely impact receipt of the complete dose—either because the device cannot deliver the complete dose, or the user "thinks" it is done and prematurely pulls the device away from their skin. These longer injection times are especially true for some biopharmaceuticals that may have (1) elevated viscosity properties when in solution, and/or (2) larger volumes within a single injection due to solubility constraints of the drug molecule. These drug solution attributes may lead to longer than expected delivery of the full dose as the patient must hold the device on the skin surface throughout the injection process. There is an upper limit that people would consider tolerable before becoming dissatisfied with how long they must keep the device on the injection site. The limit a patient will tolerate varies by individual, however, if the time required to handle the device to complete the dose is too long, sponsors should consider an alternate device (e.g. subcutaneous delivery by on-body delivery system rather than an auto-injector).

Formulation

Pain Upon Delivery

For combination products, the attention on the patient may involve the sensation of the actual delivery of the drug. This sensation is focused on the impact of the drug, not the device (i.e., poking of the needle). These sensations can be described as burning or stinging. For injectables, these may be caused by isotonicity or pH differences between the dermal area of the body and the drug solution medium. In working to stabilize the molecule shelf life, excipients may have been added to ensure the product remains active and stable throughout the shelf life but may be less comfortable when given the dose. The balance is struck through the use of buffer solutions, which provide a compatible drug formulation for patient comfort while ensuring the active molecule's purity and potency is retained.

Alternate Formulations/Delivery Devices to Meet Needs/Address Phobias

A patient-focused design for a combination product should consider how the patient will handle the product and sponsors should work to limit burdensome situations that would make the injection process an uncomfortable experience. In some cases, burdensome delivery can lead to a patient stopping their medication, all based on how the product was delivered, not an adverse reaction to the drug itself. As stated above, there could be situations where the to-be-marketed product may not have been intuitively designed for its use—or is less than ergonomically friendly. Non-patient-focused products can negatively impact successful administration of the dose and can thus lead to adverse dosing events and/or adherence problems in completing the full course of therapy.

A growing consideration for combination products delivering medications by injection are patient phobias of the needle in delivering the drug into the body. There is a balance that must be struck between patients being able to properly administer a dose by seeing where the injection is taking place versus the anxiety caused in seeing a needle being inserted under the skin. Some patients exhibit anxiety upon the mere sight of a needle, even prior to the injection. To make the dosing experience as positive as possible, sponsors have designed combination products with hidden needle features that obscure the needle. These can include designs with multiple features that shield the needle from view and access before, during, and after the injection process. Patient-focused hidden-needle designs should consider the given disease being treated carefully. This is because there may be indications (e.g., dermal conditions such as eczema) where the patient needs to be able see the exact injection site area.

Home Use

What quickly emerged from the COVID-19 pandemic was the need to continue important clinical trial studies, but where the patients could receive their medications without the requirement to travel into the clinic. The need to reduce patient exposure to the virus, along with freeing up of health-care resources, led to unprecedented change in how clinical trials were performed and resourced. This further led to the immediate need for new procedures and policies to maintain the supply of medications to patients in ongoing clinical trials that could no longer travel to the investigator's testing sites. The outcome has been the emergence of decentralized clinical trial programs, where patients could receive and administer the medication in a home setting. This process required the safe way to both (1) store and ship the medication to the patients' home, and (2) self-administration (or administration by caregiver) of the medication to ensure the complete dose was given. An important hurdle to overcome was the need to deliver medications directly to patients at their home while assuring purity and potency of the product upon delivery and then the proper administration of the medications to the patient. This process required sponsors to quickly institute procedures such as rapid delivery of investigative drug directly to the patient's home, often with the product appropriately shipped in insulated packaging. Once received by the patient, the patient needs to be able to rely on clear instructions to allow the patients/caregivers to correctly store (e.g., refrigerated vs. room temperature) product prior to taking their dose. Patient flexibility and access to care has greatly evolved post COVID-19, with a corresponding increase in the wants of patients as more convenient medical treatment becomes available. This has required the assessment that the full regimen of course of therapy was received throughout the clinical trial program.

Self-administration of drug also requires the most patient friendly delivery of the medication. Although also true for approved medications, the continuation of clinical trials during the pandemic meant that the patient had to perform any complex preparation steps of the medication, which may have been performed by a healthcare professional in the clinic prior to the pandemic. The "new normal" necessitated by the pandemic helped reinforce the need for intuitive and easy-to-use devices, which patients could easily control to successfully receive the full dose of medication. The instructions for use were validated by human factors testing, and, if needed, augmented with training (videos, phone calls, web) and follow-up monitoring to assure the dose was successfully received.

Differentiation of Device Platforms

When similar-looking devices are used for different combination products, careful design based on patient-focus research is required to minimize the risk of confusion, which could lead to improper drugs and strengths being administered. For those device platforms where the physical form is the same, clear labels and cartons, where the identifiable placement, clearly printed, of information provides the best outcomes for medication/strength differentiation. Although combination products are required to use a national drug code (NDC) to distinguish among indications, product presentations, and different medications, other means of visible differentiation are required. Another key differentiator between products can be the physical form of the device. However, the mechanics and functional requirements for delivery of drug by a multidose pen or single dose autoinjector may have physical limitations. A further differentiation feature is the use of unique colors for the device components themselves and/or highlighting information on the product label. It is important to avoid any colors that would introduce use-related risk or pose potential difficulty to users with low visual acuity due to normal aging or comorbidities due to disease (such as diabetes) or due to color-deficient vision. Differentiation of device platforms (and dose strengths of single drug within a platform) can be optimized through purposeful device platform design along with the advantageous use of color.

Clinical Trial Patient Focus for Combination Products

As discussed above, the introduction into the clinical trial program of a representative device constituent part of the to-be-marketed combination product is of high importance. The handling and use experience gained, first through the health-care professional (HCP) followed by patient/caregiver, is valuable in assuring the medication can be properly delivered for the best patient experience. This has become even more crucial with the growing trend of self-administration of medications both in clinical trials and commercial courses of therapy.

The transition from an earlier device delivery constituent part (e.g., vial/syringe) to to-be-marketed constituent part (e.g., autoinjector) through the clinical trial program may require multiple clinical studies—bioequivalence (BE), pharmacokinetic (PK), and/or other comparability bridging studies to demonstrate no adverse impact on safety/efficacy. With respect to patient-focused attributes, bridging studies concentrate on the ability of the user to successfully deliver the dose. In assessing the comparability to an earlier presentation, this includes the understanding and execution of preparation, placement on body to receive dose, activation, and timing to deliver dose and product disposal.

Combination products are designed such that formal training should not be necessary for correct use. Sponsors seek to provide the most intuitive product designs that are easily understood by both health-care professionals and patients/caregivers upon introduction of the delivery system into the clinical program. However, there may be the need to better familiarize the user with the product prior to first use in a clinical setting. This may be accomplished by a health-care provider walking the new user through the step-by-step reading of instructions for use. Once familiarized with use steps to administer the medication, the HCP/trainer may watch the patient/caregiver actually performing the dose administration. This type of introduction to the patient of new users can confirm to the HCP that the user understood and can successfully perform the dose administration steps, as well as the proper disposal of the used delivery device. Regulators may expect an evaluation of the combination product under worst case use scenarios thus training would not be acceptable for validation of the user interface when performing a summative human factors study. Certainly, what may be acceptable in a clinical setting (with HCP oversight) is in no way acceptable for the commercial presentation, and extensive human factors studies are required to demonstrate the users are able to safely operate the combination product with no external training or oversight.

The clinical trial (CT) program can provide some insight into the successful handling and understanding of the combination product. If designed to do so, the trial can be a place to gather user experience and the sponsor can learn valuable insights into how a user can use the combination product. With all CT programs, product complaints and adverse events are appropriately captured and documented, however additional user experience may be gleaned from users. Sponsors may seek to incorporate validated questionnaires within CTs to assess the patient's experiences on the ease of use and successful administration of the dose. If difficulties are observed during the CTs, improvements in the patient's experience can be addressed through further product development. These improvements may take the path of design changes to the product, but at a minimum within the instructions for use developed prior to and used in the summative (or validation) studies.

Pediatrics Considerations

In this section, we will investigate the reality of commercialized combination products intended for pediatric patients, some of the associated business challenges, as well as patient needs and device considerations.

Reality of Commercialized Combination Products

Like most things in life, generally products on the market are geared toward a broad target population, a "one-size-fits-all" approach, which meets the needs of the broadest proportion of the population. For example, a typical house meets the needs of the height of an average human, but what about those who fall significantly outside of that average? How do those who are very short or very tall cope with door frames that are too small or countertops that are too high? The same situation is relevant for most medical devices on the market that are expected to be utilized for pediatric use. But if the devices can deliver the medication to an adult, then why can it not be effective in delivering to a child? Conversely why can't a sponsor provide multiple devices to address those outside a broader target population?

The approach taken by the consumer industry and the pharmaceutical industry may not be so different, as it comes down to business costs at the end of the day. This leaves consumers or patients with two options, either adapt to the product they are using to meet their needs or pay a much higher price for a customized product—if available. In the pharmaceutical industry, most commonly, the only option for the patient is to adapt to the medical device provided as it is normally a single integral combination product, meaning drug and medical device cannot be separated. But if administered by a health-care professional or caregiver, then it should be sufficient to use on a pediatric patient, right?

Business Challenges

Historically, therapeutics, especially biologics, which are intended to be administered by subcutaneous injection, have been traditionally administered by an HCP or by a caregiver. However, requiring an HCP—or even a caregiver—to administer the medication creates a high burden for both patients and the health-care system. The benefits of self-administration in the home, both to the patient and to society, are well documented. Patients that are able to self-inject their medications gain control of their treatment schedule and setting, allowing greater independence and freedom, which may lead to better adherence and improved therapeutic outcomes.

Although it sounds like a great solution for patients, what are the reasons why there are not more custom devices intended for children to self-administer their medication at home? Unfortunately, the business case for the creation of a custom self-injection device intended for the use of the pediatric patient is a considerable challenge based on the return on investment. When putting together the business case, the challenge comes from the number of forecasted sales regarding children. Figure 3.2 shows the increased percentage use of prescription drugs as one ages and illustrating the relatively small number of children taking prescription drugs as compared to adults [8]. As the number of pediatric users requiring a self-injection device is generally much smaller than the adult population for a given medication (fortunately), developing a delivery system specifically for the pediatric users

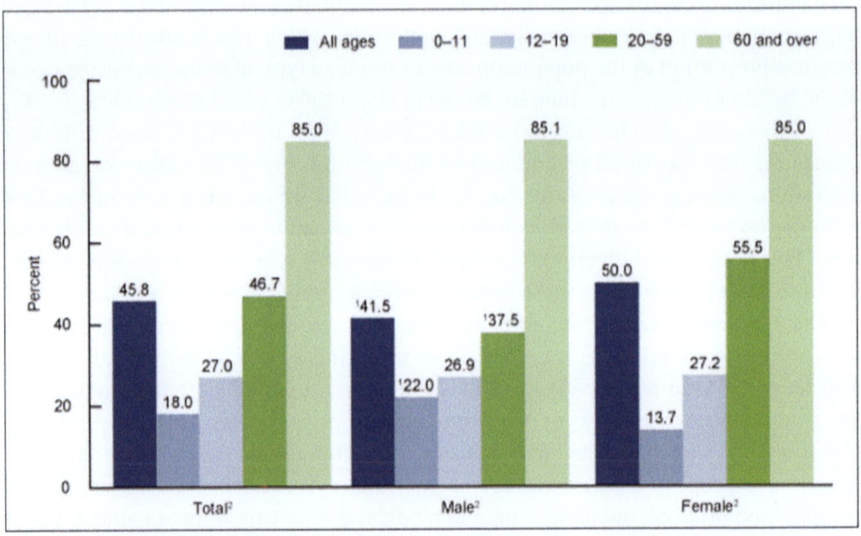

Fig. 3.2 Use of one or more prescription drugs in the past 30 days, by age and race and Hispanic origin: United States, 2015–2016

Products - Data Briefs - Number 332 - February 2019 (cdc.gov)

[1]Significantly different from non-Hispanic asian persons

[2]Significantly different from non-Hispanic black persons

[3]Significantly different from Hispanic persons

[4]Significant increasing trend with age

NOTES: Estimates for all ages were age-adjusted by the direct method to the 2000 projected U.S. population using age groups 0–11, 12–19, 20–59, and 60 and over. Access data table for Figure 2pdf icon.

SOURCE: NCHS, National Health and Nutrition Examination Survey, 2015–2016.

Chart from: www.cdc.gov/nchs/products/databriefs

represents a much more significant business challenge. The return on investment is generally not positive for a pediatric version.

Every disease will have a different proportion of adults versus children affected about which the figure above provides a general idea over time the proportion of medicine prescribed by age category. The prescriptions also include orally administered drugs, but one can deduce that pediatric injections will be the same proportion or less. As expected, children represent the lowest need and depending on the disease area, for example, oncology, which is more prevalent in adults, the business case becomes even less attractive.

The development of a combination product encompasses two main focal points around the drug and medical device, so what if the drug dose and volume is fit for both adults and children, is the device then also fit for purpose? Unfortunately, there are many other aspects to consider pertaining to the device that will be discussed in the next section. These aspects can ultimately lead to multiple devices to be developed and commercialized, which can lead to higher costs and project timing.

Differences Between Children and Adults

So, what are the differences between children and adults? Before looking at the differences, it is important to understand how different organizations globally define "pediatrics," which vary by organization as shown in the chart shown in Fig. 3.3 and the graphical representation of this in Fig. 3.4—both excerpted from ISO/DIS 23217: (N144):

There are two things that stand out when looking at the chart and the graphic above. The first being that there is no consensus for the age of adulthood, which creates difficulties depending on the region of submission. The second is a wide range of ages for pediatrics resulting in multiple physical and cognitive aspects to consider.

When defining the age of adulthood, the product sponsor must determine to which agency(ies) they plan to submit their application(s). If they plan to submit to multiple countries, the definitions from these multiple agencies will need to be selected. This presents an issue when submitting to health authorities and ultimately marketing a product for adult use as there is a significant age difference between an adult age of 22 versus 16, classified by the FDA and AAP respectively [10]. Although the difference between 16 and 22 for physiological differences (size, strength, etc.) can vary, it is not as significant as the differences in cognitive and psychosocial.

The pediatric age presents a different problem because of the amount of development and growth that occurs throughout the age range. Obviously, there are visible physical differences from the outside—but it goes further than that. When looking at the physiological, psychosocial, and cognitive differences of the age range prior to being an adult it becomes more obvious in terms of differences as outlined below. However, the challenge in developing one pediatric device becomes greater within the pediatric age range, because the device user requirements for a 6-, 10- or

Organization	Paediatric classification
Centers for Disease Control and Prevention (CDC)[23]	Infants: 0 years to 1 year Toddlers: 1 year to 3 years Preschoolers: 3 years to 5 years Middle childhood: 6 years to 11 years Young teens: 12 years to 14 years Teenagers: 15 years to 17 years
US Food and Drug Administration (FDA)[20]	Neonates: birth to 28 days of life Infants: 29 days to 2 years Children: 2 years to less than 12 years Adolescents: 12 years to 21 years(not including 22 birthday)
World Health Organization (WHO)[21]	Neonates: 0 days to 30 days Infant: 1 month to 2 years Young child: 2 years to < 6 years Child: 6 years to < 12 years Adolescent: 12 years to 18 years
Pediatric Regulation (EMA)[22]	Children: 0 years to17 years
ICH E11 (R1) [24]	Preterm infant: < 0 days Term Newborn infant: 0 days to 27 days Infant and toddler: 28 days to 23 months Children: 2 years to 11 years Adolescents: 12 years to 16-18 years
American Academy of Pediatrics (AAP)[25]	Infancy: birth to 2 years Childhood: 2 years to 11 years Early adolescents: 12 years to 14 years Middle adolescents: 15 years to 17 years Late adolescents: 18 years to 21 years

Fig. 3.3 Overview of definitions of the "pediatric population". (From Table A.1 of ISO-DIS 23217, N144)

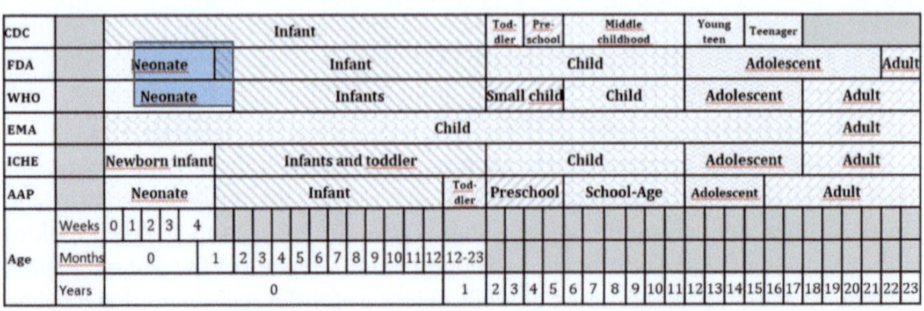

Fig. 3.4 Graphical representation of classifications of "pediatric population". (From Fig. A.1 of ISO-DIS 23217, N144)

14-year-old children will vary widely, making it very difficult to develop a device to meet the needs of all these patients.

The next section will explore in depth the importance of physiological-, psychosocial-, and cognitive-related aspects of patients to consider. This does not cover everything but gives a flavor of aspects to consider in developing the best patient-focused product for the end user.

Physiological

When looking at the anatomical differences between children and adults, it is important not to assume that children are small adults. The most obvious physical differences between a child and an adult, which are visible, include hand size, hand strength, and pinch strength, to name a few. The other not so obvious differences also play a vital role, for example, the subcutaneous thickness layer is thinner in children, proportion of fat mass is generally more in children, and muscle thickness is less in children. There is also a fundamental difference in the way children's and adults' bodies process the drug product. This may require different routes of administration, different doses, and dose regimen, and may have a very different safety profile.

These are only a few of the differences, which change as the child grows but have a significant impact on the type of injection (i.e., subcutaneous versus intramuscular), injection site, and volume delivered, to name a few. This becomes important as a thinner subcutaneous layer could lead to an intramuscular injection, causing additional pain or unintended harm.

Psychosocial

When looking at the psychosocial aspects, a child is influenced by their surroundings, which differ to that of adults and can lead to different outcomes when trying to use a device. Additionally, depending on the age of the child and their development, the way in which the device is used can vary widely. This becomes critical when products are utilized in a home setting by children, especially if they are naive to using a product and don't have parental guidance [9].

Student Orientation: Further reading and resources:(rch.org.au)

Cognitive

Cognitive abilities are obviously more developed in adults than that of children, so it is important to understand how to reach the user when interacting with a device. Jean Piaget's *Theory of Cognitive Development* (1955) determined that children depend largely on their sensory experience to learn and understand. But adults are able to use their superior cognitive ability like abstract thinking to comprehend and infer. Unfortunately, adults tend to rely too much on past experience, and assume they know how to do something but ultimately do it wrong, because the process or activity is different, this is referred to as a negative knowledge transfer as mentioned in earlier design considerations section.

Device Considerations

There are many factors to consider when designing a medical device to be used to inject an adult versus a pediatric patient as well as self-injection versus an HCP/caregiver administration. Many device options are optimized for adults; therefore, there is a risk that they do not reliably target the correct tissue layer in pediatrics or are not suitable for home use.

The previous sections touched upon some of the many patient factors to consider, but additionally, the disease indication, drug properties, and use environment must also be considered. Each of these factors contributes to the development of the device and also restricts the final design to ensure the device is useable within the defined population.

Disease Indication

The disease indication, or the disease that is being treated, can play a key role in factors to consider when developing a device for pediatrics. The user population being treated may have different physical or cognitive traits, which could lead to difficulties in using a device or treatment regime. For example, patients might have hand dexterity limitations that make it difficult to grip or hold a device or cognitive impairments, which make having too many user steps difficult to process. This will impact the size and handling of the device as well as the ease of use.

Drug Properties

The drug will require the device to handle a certain viscosity, volume, route of administration, administration site, storage temperature as well as others. This will influence, for example, the device material used, the overall device size, and the needle length/gauge. The drug's properties may influence and affect whether a pediatric user can successfully complete tasks when using the combination product. For example, high viscosity could increase the injection delivery time of the drug through the needle and increase the length of time the user must hold the injector in place to administer the drug. This could be exasperating for pediatrics due to hand strength or patience. The increase in hold time could reduce user's ability to complete an injection task and result in an underdose. The drug dosage including viscosity and volume to administer or inject will need to be considered for the physiological differences between children and adults.

In the USA, the Food and Drug Administration (FDA) may require the development of a pediatric presentation when an adult indication is submitted for approval. Specifically, the agency would require the sponsor to develop a Pediatric Study Plan or Pediatric Investigational Plan (sometimes referred to as a "PIP"). This typically involves a more complex dosing regimen than an adult indication and may require a new formulation and/or device development.

Use Environment

The administration of the drug in a hospital by an HCP provides a predictable environment to design the device around. However, as patients can—and do—self-inject in a variety of settings, under multiple different conditions, including environmental and patient induced, it becomes increasingly difficult to address all factors. The use environment for a child versus an adult also has its differences: for example, a child is supposed to self-administer or inject themselves, because adult supervision might not be always present. Children could have a higher level of embarrassment when injecting publicly or have higher anxiety around needles.

Human Factors

The focus of this section takes a look at how the patient can shape the development of the device and the importance of heeding the patients' needs when developing a patient-focused device design.

Informing Design

When designing a product in any industry, it is easy to lose sight as to who is the actual expert or the one that has the most valuable input. Is it the mechanical engineer designing the device functionality who has focused their studies and career to ensure the perfect working device? Is it the project manager ensuring a device is available for the right price who has the overarching view of everything coming together for the final device? Or is it the end user who needs to actually utilize the device to receive their medication?

The short answer is that all these aspects are important and need to be considered; however, the end user will have a significant impact on the products' acceptance if they are unable to use the combination product. A device that functions perfectly every time and is readily available at a reasonable price is worthless if the end user cannot use the product correctly.

The user needs should be used to inform the design of the device constituent part to make it intuitive for use and suit the user for any special requirements that they may have. This can create some engineering and/or budget challenges but is worth addressing early on to avoid failures during validation or worse yet after product launch in the form of customer complaints. Below is an example of user requirements for an *insulin pen-injector* and illustrates why these requirements are important.

Informing Instructions for Use

The purpose of the Instructions for Use (IFU) is to convey exactly what it says—it informs the user how to use a product. Instructions can come in many different forms including in-person training, video, verbal, hands-on, and the most basic paper format provided with the product. This section will focus on the basic paper format to understand how the user can shape the IFU.

The IFU is simple right? An engineer writes out all the tasks to be performed step by step for the user to follow. Unfortunately, a lengthy technical paper does not sit well with a user having varying degrees of literacy. Even for those who understand the language used, it can become difficult to follow along reading while trying to perform a use step and hopefully the user doesn't lose their place during mid injection as this just becomes not only frustrating but painful.

If "a picture is worth 1,000 words," it might be tempting to utilize only a single image or a few images. So why aren't IFU produced as only pictures like IKEA or LEGO instructions to simplify things and reduce packaging and translation costs? Although pictures are great, they cannot tell the whole story and, in some cases—or in some cultures—these could be misconstrued as something else leading to a misuse.

The answer is a hybrid approach in which the main steps should be demonstrated with pictures supported by brief text written in a nontechnical jargon to be understandable by a diverse literacy range [11]. The pictures can also lead to misuse if they are not interpreted correctly; so, it might be the case that multiple options may have to be tested to reduce the risk of misuse. Critical activities should be highlighted, bolded, or differentiated by color to draw attention to the user. Additional guidance can be found in ANSI/AAMI HE75:2009 (R2018), Human Factors Engineering, in which this topic is covered in some detail.

Real Life Patient Handling

Real Life Patient Handling (RLPH)—also referred to as Real Life Handling Studies (RLHSs) testing studies are not Human Factors studies. Human Factors is an established scientific discipline with set study criteria and evaluation program. RLPH or RLHS is a term used to describe any information gleaned from patient experience in interacting with the representative commercial user interface. Such information is often gathered during the continuing open-label extension studies following the Ph3 program(s). When developing a device, an engineer is primarily focused on ensuring a robust design that allows the device to function as they've envisioned it to work. However, an end user has either developed preconceptions for using a device based on previous device use or from a family member or friend, cultural influence like TV, or be completely naive in never seeing or experiencing the use of a device.

This aspect becomes critical, because everything designed into the device could be all for not if the user handles or incorrectly uses the device. When developing the device, it is important early on to obtain how the user interacts with similar devices currently on the market, complaint data for similar devices, conducting early studies to flesh out any misconceptions, and to obtain early user feedback prior to verification and validation activities.

Connected Care

Connectivity has become such a significant part of our daily life, who could imagine living without a smart phone or being connected to the internet to accomplish daily activities? Combination products associated with digital solutions can enhance their use and benefit the patient in different ways. Exploring these opportunities requires looking at the system created when a combination product is connected to an app. This broader set of components can enable a patient to manage their therapy even through digital solutions, which may not be directly linked to the combination product but will enhance the use of the combination product. In the patient's journey (as first mentioned in Chap. 1), the phases start from awareness of a symptom, through diagnosis and the receipt of treatment and their ongoing treatment, sometimes over life span. From an HCP perspective, there are different phases as well when treating a patient. These move from (1) establishing a diagnosis, to (2) giving the right treatment with (3) the right dose and frequency and finally (4) monitoring the evolution on a long-term basis. The next sections will explore how digital solutions, some themselves considered medical devices (specifically, Software as a Medical Device, or SaMD), can enhance the use of the combination products.

The patient's journey starts with the patient observing symptoms. They may not feel well but do not know if they require medical attention. This can be compounded by a patient's unwillingness to seek medical care due to anxiety being diagnosed with a disease or not being taken "seriously." This is the identification or awareness phase. Digital solutions can help patients with early identification that their symptoms could be linked to a disease requiring treatment. Use of solutions such as this could certainly reduce the delay to diagnosis and treatment. This is especially true for rare diseases (e.g., acromegaly) or diseases that are difficult to diagnose (e.g., multiple sclerosis). The digital solution could also aid in differentiation of disease (e.g., asthma versus Chronic Obstructive Pulmonary Disease (COPD)) between the first symptoms and the diagnosis that will allow the correct treatment. The challenge with such digital tools, although not directly linked to combination products, is that they can be categorized as Medical Devices. Systems classified as Software as Medical Devices (SaMDs) require the developer to follow a development process very similar to the process required for other medical devices and combination products. The intended use is key to determine the medical device classification and level of regulation required. It is important to note that a simple descriptive phrase can (even unintentionally) result in being categorized as Software as Medical Device (SaMD). For example, a symptom tracker tool—which gives a diagnosis instead of advising to seek a physician's advice—would result in the software being classified as SaMD. Given that, most of the previous chapters in which the development of combination products was addressed are relevant to SaMD, even if some differences need to be considered.

Once the patient has recognized the need for medical support, an HCP will start medical investigation to establish a diagnosis. During this phase, different digital solutions can support HCPs in understanding/interpreting the symptoms by systematic capture of parameters. These include subjective (e.g., feeling of stiffness in the morning, feeling of itching when in cold atmosphere) or objective (e.g., sensors capturing heart rate, oxygen levels). If this data can be continually collected, this data can provide valuable information on variations and trends. Typical technical solutions are websites and companion apps that (1) record patient subjective feelings, (2) capture and store time and date of medication/injection. Other solutions are off-the-shelf sensors, which capture physical activity level, heart rate, blood pressure or oxygen level. These allow for monitoring of objective parameters over longer periods of time. Other solutions are more innovative, such as proposing self-measurement of visual acuity at home, or the use of complex algorithms with machine learning and Artificial Intelligence (AI). These complex digital tools can help facilitate or confirm diagnosis for rare diseases or diseases difficult to identify with differential diagnosis (e.g., asthma/COPD, or reading of Optical Coherence Tomography (OCT) imaging).

After the diagnosis is established, the treatment can start, and digital solutions can bring a new dimension to the treatment. Monitoring the treatment (e.g., drug taken, dosage, time of dose) through digital solutions can alert and prompt a patient, caregiver, and/or HCP in the event the dose was not taken, not fully taken or was not the right drug. As an example, drugs used to reduce the risk of organ rejection after transplantation (typically immuno-suppressant medication) are particularly important to safety. Even for drugs, which are part of an ongoing or chronic therapy, monitoring these solutions can support a reduction in the dose or modifying the frequency of injection. From the patient's perspective, it can provide a feeling of control, being able to monitor some aspects of their medical condition. These digital solutions can even provide some modules to support the User Experience, for example, by providing relaxing exercise, music, or through the use of personal pictures that help patients reduce their anxiety before an injection. Besides functions that can directly support the treatment or management of disease, other support functions may be added, like reordering of drug to avoid discontinuity in treatment or connecting directly with a nurse/HCP using messaging tools to rapidly clarify questions. Connected digital tools can offer a broad range of solutions; however, consideration must be given to the complete array of components that comprise the therapy—drug, combination product, delivery system, education, patient support systems, as well as the connected health solution. The key is to avoid developing multiple independent solutions, instead of integrated solutions.

One of the most challenging stages for patients is their "first time" with the initial treatment and then maintaining the treatment over the full course of therapy. The treatment may last for years, potentially for the patient's lifetime. After all, rare is the person who WANTS to take an injection or begin the course of self-injections, which are common to most of the biologic treatments. Digital solutions can be supportive by providing digital training, from a simple step-by-step video, up to simulation tools including haptic feedback. Additional modules matching the identified pain points can be integrated as well. For example, modules for relaxation, breathing exercise, music, or app personalization can enhance patient experience and thus support persistence. A key message is to facilitate the integration of the digital solution to the patient's daily life and thus facilitate establishment of a routine that is perceived to be positive and reassuring.

There are a large variety of digital solutions that can address the needs at the various stages of the patient's journey. It is key to perform detailed analysis on specific user needs (including the patient, any caregivers, and the HCP) and identify what would best address these requirements. This is a key component to ensure adoption of the digital solution. Users do not want to have to "juggle" multiple solutions; they want to have an integrated digital solution and have the best user experience that can help them through their entire journey. To achieve this objective, it is necessary to develop integrated strategies, looking at the whole ecosystem. As described in Chap. 2, the broader ecosystem in which the patient has to manage their disease needs to be understood. This covers from the very beginning and anticipating a road map for the different functions to be implemented over time, or as the needs are changing. This road map should consider the solutions being complementary. The road map ensures commonalities, the use of different communication channels (apps, SMS, messaging, social media, web page). It must aim to provide a sense of caring, providing an integrated and positive patient/user experience. Finally, and perhaps most critically, the solution provided must address the confidentiality and use of the data collected. Trust needs to be established during the first steps of use of the digital solution. When looking at the drop rate when using a new app, it is clear that the patient providing private information (e.g., name, birth date, mail address) must have trust that their personal confidential information (or "PCI") is protected throughout all parts of any connected solution. This is absolutely critical in adopting use of the digital tool and can result in abandoning using the app. It is to note that failure to provide a secure channel, which ensures confidentiality of PCI and complies with the legislation of the country of use, can have significant legal and financial implications to the product sponsor.

The underlying power of digital solutions is the collection of a large set of data. The collection of this data can be done on-site or remotely and captured in real time or deferred for later analysis. The use of newly accessible technology like Artificial Intelligence or Machine Learning is able to rapidly treat and analyze the data in a meaningful way. Developing algorithms to analyze the data may be necessary; however, the quality of the data collected is vital to efficient use of the digital capabilities. Often data is collected with little anticipation of how the data will be used, or in what format it should be collected. Even more critical, the data transfer and storage must be set up and controlled to ensure the integrity of the data to allow it to be analyzed in a way that is useful. This requires regular updates to maintain security of the data; however, the cost of data usage and storage over time is often fully ignored. When developing digital solutions, it is key to consider the data strategy from the start. This is not only key, because it impacts the cost, timeline, and development strategy, but because the business value of the data is often neglected. Digital solutions can play a role in facilitating access to medicine, maintaining adherence over time, and could well be required to support reimbursement by the payer.

Digital solutions can greatly enhance the patient's experience; however, the regulatory classification makes integration of different solutions in an ecosystem quite a challenge. Each element of the ecosystem can vary from nonregulated to regulated. Most SaMD regulations and classification guidance have been issued over the last 5 years, with no harmonized approach worldwide. U.S. and EU regulations may classify the same digital solution differently—the same system can be nonregulated in the USA, while it could be deemed as regulated in EU. It is important to check for each region whether additional specific regulations need to be considered.

The integration of different elements of the ecosystem can be challenging as well in cases where one of the elements of the digital solution is considered SaMD. Currently, the regulators consider that if one element of the system is SaMD, then the full ecosystem, even including non-SaMD elements, will still be considered, in its entirety, as SaMD. A growing concern for the regulatory classification is the level of feedback of the SaMD to enhance patience experience. As more combination products contain embedded sensors, the level of validation data for the digital readouts to patients of a successful dose received will be a growing challenge.

Many questions remain in the integration of digital solutions for combination product pharmaceuticals. In the future, will there be some possibility to consider some elements as a "black box," while other elements are regulated? This is a question that may need to be addressed in the future in order to allow pharma to fully enter into the digital space. Will the Health Authority accept applying the SaMD development process only to the relevant modules while performing the validation exercise on a complete system? Another challenge relates to the nature of software, which is permanently evolving, either to include security patches, update of the operating system, or hardware change. The opportunities offered by all new standard functions (e.g., better camera, integrated double security) require system agility with rapid versioning updates. Will the regulator provide a path forward to facilitate update and improvement of SaMD solutions without activating the heavy and somehow slow regulated path? All these questions are key to providing integrated digital solutions that can evolve rapidly, thus following the accelerated pace of digital development.

Conclusions

The opportunity to incorporate patient-focused design should be emphasized early in the development of the combination product. User needs in delivery of the drug should be at the top of the sponsor's development plans and efforts in the design and evaluation of the combination product. User requirements should encompass the entire user groups that include unique patient attributes (physiological, psychosocial, and cognitive) of any patient that may use the combination product—including, if applicable, pediatric patients. The development plan needs to consider the entire user interface: device handling, drug delivery, product packaging, instructions for use, digital solutions, embedded sensors, data analysis, and user apps.

The focus on the patient and their needs has led to treatment and user interface preferences, which may favorably impact patient's perceptions in taking their medication, and potentially improve patient adherence/persistence to treatment. Patients who have a satisfying experience taking their medication will be better able to manage their disease. Perceptions of onerous treatment schedules and/or unfriendly devices can, however, increase patient frustration or anxiety, causing patients to see less benefit and more bother to continue taking their medication. The large number and variety of digital solutions highlights the opportunities that exist in addressing patient needs throughout all stages of the patient treatment journey. The benefits of these digital tools are increasingly being used—from early diagnosis all the way to the end of therapy. All of these patient-focused considerations in design development plans for combination products provide the best likelihood for favorable patient perceptions and positives actions in their treatment leading to an effective impact on health outcomes.

References

1. Patient perceptions of injection devise used with dulaglutide and liraglutide for treatment of type 2 diabetes. Matza, et.al. Current Medical Research and Opinion, 34:8, 1457-1464 (2018)
2. Hold the device against the skin: the impact of injection duration on user's force for handheld autoinjectors. Andreas Schneider, Philippe Mueller, Christoph Jordi, Philipp Richard, Peter Sneeringer, Ranjan Nayyar, Mary Yovanoff, Jakob Lange, Expert Opinion on Drug Delivery, https://doi.org/10.1080/17425247.2020.1704730, https://pubmed.ncbi.nlm.nih.gov/31835921/

3. FDA Draft Guidance for Industry, Essential Drug Delivery Outputs for Devices Intended to Deliver Drugs and Biological Products. June 2024
4. Guidance for Industry, Food and Drug Administration Staff and Other Stakeholders. Patient-focused Drug Development: Collecting Comprehensive and Representative Input. June 2020
5. Guidance for Industry, Food and Drug Administration Staff and Other Stakeholders. Patient-focused Drug Development: Methods to Identify What Is Important to Patients. Feb 2022
6. Guidance for Industry, Food and Drug Administration Staff, and Other Stakeholders. Patient-focused Drug Development: Selecting, Developing, or Modifying Fit-for-Purpose Clinical Outcome Assessments. June 2022
7. Guidance for Industry, Food and Drug Administration Staff, and Other Stakeholders. Patient-focused Drug Development: Incorporating Clinical Outcome Assessments into Endpoints for Regulatory Decision Making. Draft April 2023
8. Products-Data Briefs. Number 332 February 2019. https://www.cdc.gov/nchs/products/databriefs
9. Student Orientation. Further reading and resources, https://www.rch.org.au/studentorientation/further reading and resources/
10. Draft ISO standard ISO DIS 23217.2:2023 – *"Injection Systems for self-administration by paediatric patients – guidelines for design"*
11. Guidance for Industry and FDA Staff: Application of Human Factors Engineering Principles for Combination Products: Questions and Answers. Sept 2023

Chapter 4
Patient Focus in Drug Delivery Under Real World Constraints

Sherri Biondi, Joanna So, and Rachel Poker

Abstract Pharmaceutical development teams are highly cross-functional and bring together a broad range of disciplines with a goal of developing a safe, effective, usable, and manufacturable combination product. During development, the needs of the different team members become apparent and can sometimes seem in conflict with the ideal patient-centric product presentation. In this chapter, we will explore these different needs and discuss how they might impact patient-centric decision making. In some cases, the usability engineer will need to consider their design options recognizing the constraints of the molecule itself and how it needs to be delivered. In other cases, the engineer will need to make trade-offs considering business constraints. Different examples will be discussed that highlight options for compromise and finding a middle ground. Ultimately, the earlier that developers can anticipate and understand product-specific constraints and trade-offs, the better equipped they are to make the best design decisions for their organization and for their patients.

Keywords Constraints · Trade-offs · Drug product needs · Device needs · Business needs · User needs

S. Biondi (✉) · J. So (✉)
Device Development, Dosage Form Design and Development, Biopharmaceuticals R&D, AstraZeneca, South San Francisco, CA, USA
e-mail: sherri.biondi@astrazeneca.com; joanna.so@astrazeneca.com

R. Poker (✉)
Pharmaceutical Development & Manufacturing, Gilead Sciences, Inc., Foster City, CA, USA
e-mail: rachel.poker@gilead.com

R. R. Nesbitt et al. (eds.), *Patient-focused Development of Biopharmaceutical Products*, AAPS Advances in the Pharmaceutical Sciences Series 27,
https://doi.org/10.1007/978-3-031-98253-8_4

Introduction

The previous chapters in this book have explored what it means to be patient-focused developers. This chapter is different, because it centers on the reality that product developers often must accept balancing patient-focused design decisions with other product requirements. Put another way, patient needs are just one of the many considerations that are competing for prioritization during biopharmaceutical product development. One development path may be the obvious choice to best optimize for patient centricity, but there are often constraints or conflicting requirements that result from the nature of the drug itself and how it must be delivered as well as business constraints such as timeline, cost, and manufacturability. Developers often need to balance design and business requirements with user requirements to best optimize functional reliability, robustness, usability, and the business constraints for developing and launching a product. It is also important to acknowledge that patients are diverse and have correspondingly diverse preferences. It may not be meaningful to reduce the preferences of different patients to an average or percentage in favor of one solution, calling that solution "patient-preferred." Sometimes, catering to a preference for one aspect of a product may impact the ability to deliver on a preference for a different aspect of that product.

The question then becomes, how does an organization effectively implement patient centricity in the real world? What are the other criteria that influence decision making in drug development, and how does an organization balance these differing needs to successfully commercialize a patient-focused product?

Constraints and Trade-Offs

Biopharmaceutical product development always has constraints and trade-offs. Product developers can aspire for the most user-centric product possible; however, they must acknowledge the constraints put in place by the nature of the therapeutic and the business. Figure 4.1 illustrates examples of the constraints and trade-offs that might be considered among the drug product, the device, the business needs, and the user needs. These will be discussed in more detail throughout the chapter.

In prior chapters, user needs were discussed generally. Different user populations use injection devices (e.g., health-care providers, caregivers, and patients), and each group has needs that can generally be summed up by saying that: (1) the product needs to be understandable to each user group so that they know how to use it safely and effectively, and (2) each of these user groups also needs to be physically capable of using the product safely and effectively. These two general needs create a series of user requirements that guide product design and development. Let's consider how these balance with the different constraints described above.

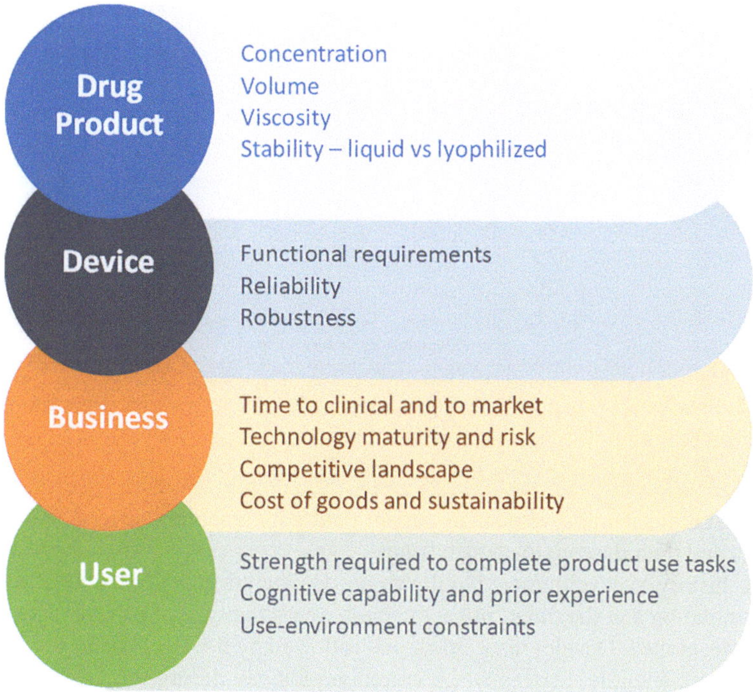

Fig. 4.1 Example of constraints and trade-offs

Drug Product Needs and Related Constraints

Drug products (the molecule itself and the needed formulation) have many requirements to achieve the therapeutic dose and safety profiles. These drug product needs often place constraints on product development, with requirements such as dose strength and constraints such as stability requirements. With this in mind, device developers need to consider that drug products might be liquid or may need to be lyophilized. They may be low or high concentration; perhaps low or high liquid volume; and the resulting viscosity has the potential to impact the user's ability to deliver the product easily. Each of these impacts the product design and patient-centric design decisions.

As an example, consider the drug product constraints of different biologics that have been developed over the last 25 years. Enbrel® was early to market in this space to treat rheumatoid arthritis (1998, Fig. 4.2) and was initially commercialized as a lyophilized powder in a vial. Since the preparation of this product might be complex for a layperson, additional product presentations were developed.

Enbrel was later reformulated into a liquid drug product. With that flexibility multiple product presentations were subsequently developed and commercialized including liquid vial, prefilled syringe, single-use mechanical autoinjector, and reusable electromechanical autoinjector (Fig. 4.2). These product presentations

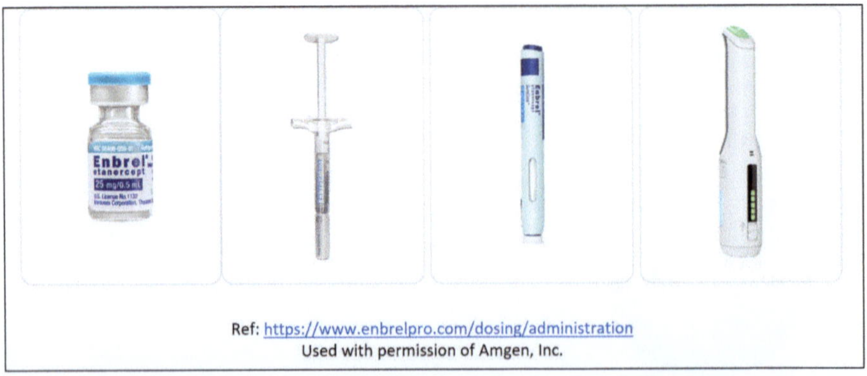

Fig. 4.2 Breadth of Enbrel product presentations. (Ref: https://www.enbrelpro.com/dosing/administration. Used with permission of Amgen, Inc.)

were a significant improvement in patient centricity, enabling several options for patients and their doctors to choose from.

This history of development highlights the constraints imposed by the drug product formulation and the direct impact it can have on the type of product that is ultimately developed. Development into a vial followed by life-cycle management into more patient-friendly devices was a common pathway during the earlier years of biopharmaceutical drug development.

The success of these early products resulted in an industry-wide shift toward patient-centric devices that could enable self-administration. In recent years, manufacturers have worked to launch their products in these presentations—developing the formulation and device early with a user-centric commercial product in mind. For injection products used to treat chronic diseases, this typically means trying to develop a liquid formulation at a concentration sufficiently high that the dose volume is less than 2 mL, fitting within the standard format of existing, mature prefilled syringe and autoinjector platforms. Development effort is also put into minimizing the drug product viscosity to enable delivery within the strength capability of the user population. Many prefilled syringes and autoinjector products have been commercialized in recent years (Table 4.1) and generally provide a convenient and intuitive way for people to self-administer therapies critical to managing their disease.

Some drug products are less accommodating than the products described above and might face delivery challenges that require less frequently used device technologies. For example, some molecules are not stable in a liquid presentation and might need to be lyophilized. Some simply require a large dose or cannot be concentrated to enable a low delivery volume; these require devices that can deliver larger volumes. For these two situations, developers often consider the risks and benefits of different development and delivery options and discuss these with their teams. They might consider the following:

Table 4.1 Recently approved prefilled syringe and autoinjector products

Brand	Molecule	Manufacturer	Device	Approved
MAKENA (withdrawn)	hydroxyprogesterone caproate	AMAG	Autoinjector	2018
ILUMYA	tildrakizumab	Sun / Merck	Prefilled syringe	2018
KEVZARA	sarilumab	Sanofi / Regeneron	Autoinjector	2018
AIMOVIG	erenumab	Amgen	Autoinjector	2018
PALYNZIQ	pegvaliase	BioMarin	Prefilled syringe	2018
FULPHILA	pegfilgrastim	Mylan	Prefilled syringe	2018
SUMATRIPTAN	sumatriptan	Golden State Medical Supply Inc	Autoinjector	2018
ATROPINE	atropine	Rafa	Autoinjector	2018
NIVESTYM	filgrastim	Pfizer	Prefilled syringe	2018
EPINEPHRINE	epinephrine	Teva	Autoinjector	2018
EMGALITY	galcanezumab-gnlm	Eli Lilly	Autoinjector	2018
XYOSTED	testosterone	Antares	Autoinjector	2018
TEGSEDI	inotersen	Akcea	Prefilled syringe	2018
HYRIMOZ	adalimumab	Novartis Sandoz	Prefilled syringe	2018
HYRIMOZ	adalimumab	Novartis Sandoz	Autoinjector	2018
UDENYCA	pegfilgrastim	Coherus	Prefilled syringe	2018
UDENYCA	pegfilgrastim	Coherus	Autoinjector	2018
ACTEMRA	tocilizumab	Roche (Chugai)	Autoinjector	2018
TREMFYA	guselkumab	Janssen	Autoinjector	2019
SKYRIZI	risankizumab	AbbVie	Prefilled syringe	2019
COPAXONE (OUS only)	glatiramer	Teva	Autoinjector	2019
NUCALA	mepolizumab	GlaxoSmithKline	Prefilled syringe	2019
NUCALA	mepolizumab	GlaxoSmithKline	Autoinjector	2019
VYLEESI	bremelanotide	Palantin / AMAG	Autoinjector	2019
HADLIMA	adalimumab	Merck / Samsung	Prefilled syringe	2019
HADLIMA	adalimumab	Merck / Samsung	Autoinjector	2019
GVOKE	glucagon	Xeris	Autoinjector	2019
TERIBONE (OUS only)	teriparatide	Asahi Kasei	Autoinjector	2019
XOLAIR	omalizumab	Genentech	Prefilled syringe	2019
FASENRA	benralizumab	AstraZeneca	Autoinjector	2019

(continued)

Table 4.1 (continued)

Brand	Molecule	Manufacturer	Device	Approved
ZIEXTENZO	filgrastim-sndz	Novartis Sandoz	Prefilled syringe	2019
ABRILADA	adalimumab	Pfizer	Autoinjector	2019
AJOVY	fremanezumab	Teva	Autoinjector	2020
DUPIXENT	dupilumab	Sanofi	Autoinjector	2020
HULIO	adalimumab	Mylan	Prefilled syringe	2020
HULIO	adalimumab	Mylan	Autoinjector	2020
IMRALDI (OUS only)	adalimumab	Biogen / Samsung	Autoinjector	2020
ZEGALOGUE	dasiglucagon	Zealand	Autoinjector	2021
SKYRIZI	risankizumab	AbbVie	Autoinjector	2021
WEGOVY	semaglutide	Novo Nordisk	Autoinjector	2021
TEZSPIRE	tezepelumab	AstraZeneca/Amgen	Prefilled syringe	2021
NALOXONE (discontinued)	naloxone	Kaléo	Autoinjector	2022
MOUNJARO	tirzepatide	Eli Lilly	Autoinjector	2022
MIDAZOLAM (discontinued)	midazolam	Rafa	Autoinjector	2022
IDACIO	adalimumab	Merck / Frezenius	Autoinjector	2022
TEZSPIRE	tezepelumab	AstraZeneca/Amgen	Autoinjector	2023
COSENTYX	secukinumab	Novartis	Autoinjector	2023
CYLTEZO	adalimumab	Boehringer Ingleheim	Autoinjector	2023
YUFLYMA	adalimumab	Celltrion	Prefilled syringe	2023
YUFLYMA	adalimumab	Celltrion	Autoinjector	2023

The list is not exhaustive

Considerations for Lyophilized Drug Products

- The lyophilized drug could be developed and commercialized in a vial, or it could be developed in a delivery device with reconstitution capability.
- If the product is stored in a vial, developers will need to consider the user's ability to perform the reconstitution, syringe filling, and dose setting steps and how the product might be labeled for this to be intuitive and minimize use-errors.
- Developers need to recognize that reconstitution adds complexity to dose preparation tasks and will need to decide if administration should be restricted to HCPs in-clinic to ensure safe and effective administration.
- If developers pursue self-administration by patients at-home (offering convenience and autonomy for patients), there may be added risk of dosing errors with the extra steps and materials needed during dose preparation.

- Developers will need to consider if the diluent should come in a vial or a prefilled syringe. A prefilled syringe can reduce the number of use steps and the potential for dosing errors.
- Developers will also need to consider whether preparation and administration components such as diluents and syringes are offered together with the drug or as separately obtained components. Having components packaged together can increase convenience and reduce error for patients but can increase manufacturing costs and supply risks.
- Development of the drug in a delivery device with reconstitution capability is an alternate option and can offer patients an at-home all-in-one solution, but the more advanced technology carries more development and business risks—the consequences of which are potentially a delay in bringing the product to market and to patients in need. Additionally, ensuring the reconstitution device is intuitive and usable for patients is not trivial.

Each of the options mentioned here involves different trade-offs, with different risks and impact on patient use and product development timelines. The earlier these decisions can be considered, the more likely a patient-centric decision can be pursued.

Considerations for High Dose and Large Volume Delivery

Developing a patient-centered delivery device for high dose and large volume delivery can be challenging from a usability perspective. One choice for developers might be to concentrate their drug products to keep the delivery volumes as low as possible. These higher concentration formulations can be very viscous, complicating the delivery experience. If a manual device is developed, the strength capability of the intended user must be considered to ensure they can deliver the drug comfortably and successfully, particularly if the user group has any dexterity limitations. If an automatic delivery device such as an autoinjector is considered, developers need to consider use-related risks related to long delivery times and the potential for incomplete dosing.

A second option might be to choose a lower concentration and lower viscosity dosage form. This increases dose volume and requires developers to consider how that larger volume will be delivered. There are several strategies in these situations—each with their pros and cons from a patient centricity perspective:

- The therapeutic could be delivered in a clinical environment where large volume delivery can be managed safely and effectively by a health-care provider. This can be inconvenient for the patient and may impact adherence to therapy if scheduling is too challenging.
- The therapeutic could be delivered in a large volume on-body delivery system (OBDS). These have been commercialized for a small number of therapeutics (Fig. 4.3: Furoscix, Skyrizi) and are sufficiently intuitive to enable

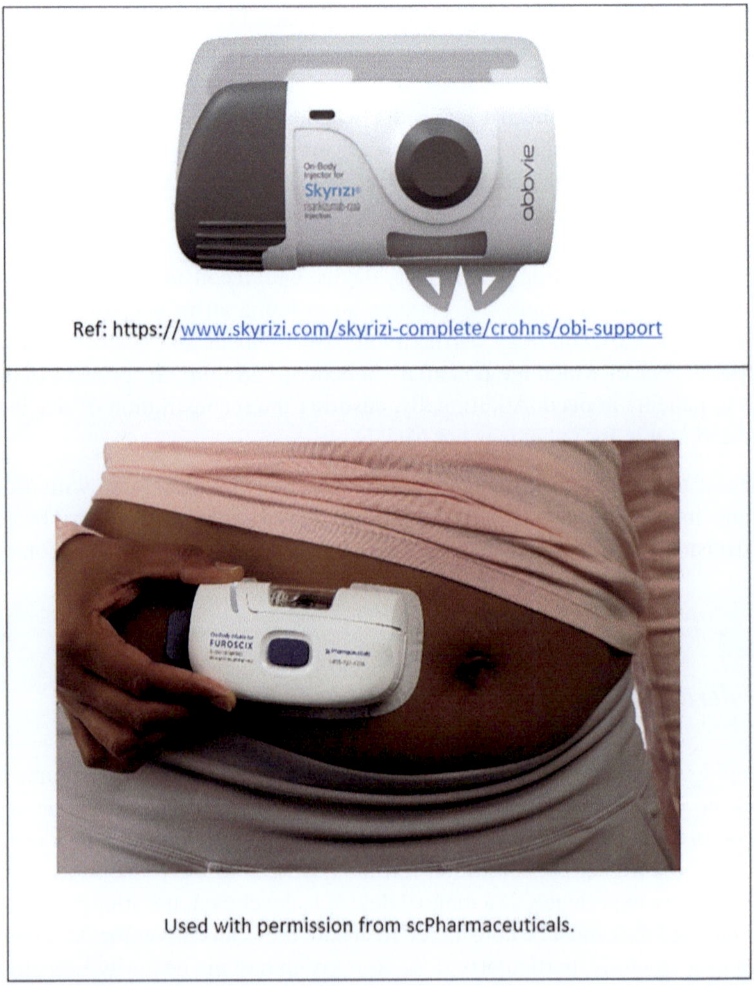

Fig. 4.3 On-body injector devices. (Ref: https://www.skyrizi.com/skyrizi-complete/crohns/obi-support. Used with permission from scPharmaceuticals)

self-administration. That said, these products have more preparation tasks than a prefilled syringe or autoinjector and longer delivery times, which might present challenges for some users. Additionally, the developer needs to consider whether the patient population will have wearability concerns (application, administration time, removal) with the OBDS.

- Another choice might be to increase the dosing frequency. For example, if a drug product is formulated into an 8 mL dose volume that needs to be delivered monthly, a developer might consider delivering the dose in a single large volume OBDS or they might increase the dosing frequency and deliver ~2 mL in an autoinjector weekly (depending on the pharmacokinetic (PK) profile). Some patients might prefer the less frequent dosing at the expense of a longer dosing time (potentially tens of minutes), while other patients might prefer more frequent dosing if the delivery time is in seconds.

The examples described here (lyophilized dosage form and high dose delivery) are just two examples where the constraints of the drug product have a potential to impact the patient experience and the usability of the product. There are typically many ways to address these challenges in the product design. The key is to consider these early to ensure that development is on the path to the desired commercial product.

Constraints and Trade-Offs Between Device Functional Robustness and Usability

In the design of multicomponent drug delivery devices, we are often asking the device to do many things—be functionally reliable, be robust to expected (and unexpected) use scenarios, and be intuitive and easy to use. Due to the abundance and complexity of design requirements that must be satisfied, it is not uncommon to encounter competing requirements for functionality and usability, where trade-offs must be carefully considered, and compromises must be made.

Combination products have strict functional requirements that ensure the product operates safely and effectively. Many of these performance requirements involve actions that the end-user must carry out in order to properly receive their dose. For example, in order to receive a dose from a prefilled syringe, a self-administering patient must remove the needle cap, insert the needle, and push down on the plunger rod. Correspondingly, there will be design requirements for the maximum force required to remove the needle cap and for the maximum force required to push down on the plunger rod. As noted above, it is important that the forces needed to operate the device are achievable by the intended user population. Interestingly, requirements such as "needle cap removal force" are often more complex than some might initially think. As is often the case, there are competing considerations. In addition to the removal force being within the user's strength capabilities, the cap must also be held tightly enough to maintain sterility throughout all shipping and handling. This second requirement imposes a limit to how easily the cap can be removed without risking sterility of the product. When these two competing requirements are considered together, developers might realize it is difficult to meet both fully. As a result, the needle cap may need to be more difficult to remove than some lower-strength users might prefer. In the extreme, there may be cases where a small proportion of the weakest users (e.g., those with high degrees of dexterity impairment) find the cap removal challenging. Design affordances that promote easier cap removal, such as large grip features or the ability to rotate the cap to remove, can help bridge these competing requirements. If these design features are a necessity for a particular patient population, they should be considered early in development to ensure on-time implementation.

Autoinjectors provide another example where functionality and usability considerations must be balanced. With push-on-skin autoinjectors, the injection is typically initiated after the user pushes the device against the body. A developer might consider whether users have the physical strength to activate the device. Notably, not pressing hard enough to activate the injection is a known use issue for autoinjectors, so keeping the force reasonably low can help patients more intuitively and confidently use the device. At the same time, developers must consider representative handling of the device and whether there is a risk of unintentional activation, for example, if someone prematurely applies pressure to the autoinjector needle shield. To mitigate this risk, a developer might want to increase the force required to activate the device. Developers must consider how to balance the need to keep forces low enough to activate the injection when needed, while high enough to avoid unintentional activation.

Across all device types, developers will find that they have to manage conflicting requirements such as these. Design affordances should be considered where possible. In the end, design decisions will need to balance the spectrum of functionality and usability needs as well as reasonably possible.

Business Needs and Related Constraints

Developers are often challenged by timeline and cost constraints during development, especially if they might impact launch timelines or the possibility to launch in certain cost-constrained global regions. When considering timelines, the cross-functional team must consider the appropriate start-date considering the maturity of the formulation and device technologies, the manufacturing process, and the development risks the team will face. When considering product cost, developers might realize that certain device features are desired but add significantly to the product's cost of goods, resulting in a product so expensive that it cannot be marketed across the desired global regions. Capability to manufacture and reliable supply are critical considerations as well. All of these elements are ever-present realities in product development, and design choices need to be mindful of these constraints up-front.

Speed to Market

Timelines for new technology development can be long and uncertain, sometimes taking multiple years to have a functionally reliable and robust product. To deliver a new technology on the timeline needed by a molecule team, it is typically necessary to start early and have a reasonably mature technology at the time it is married with a molecule's development program. If this is not possible, teams will need to debate whether selecting an immature new technology with potential schedule delays is better than selecting a more established technology with a faster and more predictable time to launch. Sometimes the more patient-centric decision might be prioritizing drug access and speed to market. There are many examples, like Enbrel (described earlier), where patient-centric design implementation has been stepwise and implemented as life-cycle management, over years.

Cost and Patient Access

All products need to be manufactured within the cost constraints of the markets they will be distributed in. Across the globe, some markets have more challenging cost constraints than others. With this in mind, product cost constraints (materials and manufacturing costs) might limit device design features in order to enable patient access across the broadest range of markets. An expensive drug delivery device relative to the allowable sales price of the product can result in a drug no longer being commercially viable in certain regions of the world, limiting patient access to beneficial therapies. Manufacturers may need to weigh the trade-offs of providing a drug to patients in low-cost, lower-featured devices versus not bringing the product to market at all. Having a patient focus in these cases may once again mean prioritizing drug access over integration of high-cost technology.

Manufacturability and Supply Continuity

Device technology decisions are often made early in development, when limited information is available regarding manufacturability and supply-related risks. A developer might not think about how certain use-related device features may create challenges in manufacturing. When organizations consider the trade-offs between device features, it is important to include the perspective of the manufacturing and supply groups so that associated risks can be considered and mitigated early. For example, some novel device types might require new manufacturing facilities to be built or significant changes to existing facilities. This can be time consuming and add significant cost to the product beyond the component costs from the device itself. A second consideration is also supply risk. Some device technologies require novel components that may be single-sourced and unavailable in the required volumes. If those components are not consistently available, product manufacturing will not meet demand, and patients might not have access to the needed medicines. Both manufacturability and supply risks should be considered early as device technologies and device feature decisions are being made so that trade-offs and constraints can be considered in a balanced way—with all aspects of the product in mind.

Trade-Offs Vetted and Decisions Made for Best Real-World Use

With all the challenges and available options described above, how are decisions made leading to the to-be-marketed product? How do the drug and device experts come together to balance and hone priorities down to a single development path? It takes great coordination and education as the product team brings clinicians, chemists, marketers, and engineers together in patient-focused decision making. It's not just bringing these diverse groups together into the same room but having learning sessions that explain all aspects of the drug/device product. A clinician can buy into instructions for use language if they understand why the device would not deliver drug as intended based on the latest human factors testing. Alternatively, an engineer can appreciate the need for a design change based on drug formulation or storage conditions. These two examples illustrate that issues will emerge, and trade-offs need to be made continuously throughout the development process—from early drug plans to the final user interface packaging. These priority discussions can be quite intense as timelines and development costs are vetted with the patient treatment needs and product design considerations. The final decisions, however, are the most critical (and rewarding) in setting the product development path leading toward a more beneficial/effective real-world patient-focused product.

Conclusion

Drug product and device developers have a number of considerations to balance as they plan the product presentation. Beyond the challenges of clinical development and the needs to meet all the regulatory requirements, the product sponsors need to trade-off various factors that will impact their product presentations, the timeline and costs of development, approval path, and, ultimately, the patients' ability to use the product as intended. Pipeline molecules across industry have a broad range of needs to deliver a safe and efficacious dose. These needs often require device developers to make compromises to the simplest and most intuitive product design that patients would prefer. When teams identify these needs early, developers can still maintain a patient-focused mindset by considering how they can meet patient needs within these constraints. Guiding design principles should be considered around simplifying workflow and ensuring dimensions and forces are within user capability. Early cross-functional partnerships and an early start to development can provide the best opportunity for novel patient-centered technologies to come to market. The earlier that developers can anticipate and understand the constraints and trade-offs, the better equipped they are to make the best design decisions for their organization and for their patients.

Chapter 5
The Commercial Viewpoint

Michelle R Blechman and Colleen Foley

Abstract This chapter provides a deep dive into a key element of successful commercialization of new devices and delivery systems: understanding the end users. It demonstrates the need to incorporate deep insights of patients and caregivers at various stages, ensuring that both the design and rollout phases address relevant human attitudes and behaviors. This chapter also defines best practices for obtaining those insights appropriately, and sufficiently early in the product development process—describing the role of System 1 and System 2 thinking in the adult decision-making process as it applies to adopting new delivery devices. It also provides some learnings gleaned from case studies in device/administration system development. Finally, it includes some guidelines—the Top Ten Truths—plus resources for validating end user decision-making and adoption.

Keywords Insights · Behaviors · Attitudes · Decision-making

This chapter provides a deeper dive into achieving patient centricity both earlier in the development timeline and throughout the commercialization process, using tools and resources that help allow developers understand human issues affecting the adoption and appropriate use of new devices or delivery systems.

This chapter focuses on adult decision-making. This focus is important, because for many illnesses and needed treatments, patients are diagnosed as adults and, therefore, must "learn" a new habit to help them live with and treat their condition.

This chapter will also focus on product usage information that may be learned too late in the development process, and steps on how to avoid that trap. Finally, this chapter will provide some general guidelines from multiple experiences in

M. R. Blechman (✉)
Brandirections LLC, Deerfield, IL, USA

C. Foley (✉)
Marketing Research/Commercial Insight, Abbvie and Horizon (Amgen) Therapeutics, Chicago, IL, USA

R. R. Nesbitt et al. (eds.), *Patient-focused Development of Biopharmaceutical Products*, AAPS Advances in the Pharmaceutical Sciences Series 27,
https://doi.org/10.1007/978-3-031-98253-8_5

device administration system development. At the end of the chapter, we will share what we call

1. The Top Ten truths
2. Resources for validating patient decision-making and adoption

As discussed in Chap. 2, patient *ecosystems*—the *enablers* for the patient—can help or hurt outcomes. Also critical to outcomes are patient/caregiver assumptions about the benefits of the therapy; if their assumptions are not realistic, adults may be unwilling to adapt or change their behavior to get the optimal outcomes from the therapy. Every aspect of device/administration systems needs to consistently identify and address the patient's and/or provider's viewpoint in the actual handling and administering and, importantly, in how the device fits into their beliefs about the condition and their lives.

Neurological Study and Decision-Making

The neuroscience of the brain enables humans to make very quick decisions for habitual, repetitive behaviors and can be referred to as System 1 thinking. Decisions that require more thought, information, and consideration are referred to as System 2; human behavior tends to minimize the number of times System 2 thinking is engaged. Uncovering these decision-making insights and dynamics requires rigor. The methods for doing so are not necessarily complicated but need to be planned for in the project timeline and budget.

The importance of a comprehensive view of decision-making becomes clear when a product is introduced to the patient, accepted, and then, in the trial period, rejected and never refilled. As previously discussed, if the product is used incorrectly and the patient is aware of the problem, it then requires replacement. If the patient is unaware, they may perceive a lack of efficacy due to misuse, but the misuse is never identified by the patient or their doctor. The resulting inefficacy is misattributed to the product rather than poor user experience.

Fully explored and explained in Daniel Kahneman's *Thinking, Fast and Slow,* neurological study asserts that judgment and choice involve two systems named System 1 and System 2. This approach draws on cognitive and social psychology.

System 1

Operates quickly, automatically, with little/no effort and no sense of voluntary control. It's intuitive judgment, based on bias and heuristics and recognition of cues and associations. It involves emotions and memories—the Remembering self. It allows humans to have a way to confront problems with ease and comfort: that is, I know what this is; I have seen this before. According to Kahneman: "The intuitive System

1 is more influential than your experience tells you and it is the secret author of many of the choices and judgments you make." *(ref: Thinking Fast and Slow, Daniel Kahneman, Farrar, Straus and Giroux, 2022, p13).*

Most of what we think and do originates in System 1. System 1 effortlessly originates impressions and feelings that are the sources for the beliefs and choices of System 2.

System 2

Allocates attention to effortful activities; it's controlled operations and harder to do. When an intuitive solution fails, or when an unfamiliar set of circumstances is present, we will engage System 2. It is the system that describes how we think of ourselves: a conscious reasoning person who has beliefs, makes choices, and decides what to think and do—the Experiencing self. It includes diverse operations that require attention. System 2 can inform automatic functions of attention and memory and has some ability to change System 1. System 2 takes over when things get difficult.

So, is it surprising then, with this level of judgment/decision-making complexity that humans engage in, that we cannot risk making assumptions about how our prospective users will receive our ideas/solutions? Our own System 1 is potentially leading us to incorrect assumptions.

How to Avoid the Traps?

The simple answer is knowing your *intended users* and how they process and engage with their condition and treatments. Uncovering these insights historically has been a responsibility of the commercial organization when a product is in late-stage development (Phase 2B or early Phase 3). Given that the development team and the commercial team have similar needs to know their audiences in depth, early coordination in these efforts seems merited. Each discipline needs to appropriately gather and synthesize the insights. The recommended steps are outlined as follows:

- *Analysis Phase:* The development team should map out the current state of treatment/administration and identify gaps. Work with your market research organization and anyone in the commercial organization who knows the therapy area.
  - Identify real and perceived competition, including another treatment, or no treatment, or workarounds.
  - Identify where in the patients' journey current solutions reside.
  - Determine the likely success of current options in terms of adoption, efficacy, uses, and reuse. This evaluation is achieved through activities that expose solutions and designs to the prospective patients. Note that while you may

want to understand *stated* patient preference for designs/solutions, you need to understand deeper, less obvious biases influencing your prospective patients. In a structured preference study, a patient can pick out what they prefer, but often it will not guarantee their adoption of their "preference" later.

The mapping process forms the basis of the patient journey.

As mentioned in Chap. 1, the patient's journey is a critically important step in the process, which:

1. Documents common paths for patients - from symptom recognition through diagnosis, treatment and treatment changes, and other life events with the condition.
2. Identifies missed opportunities for helping patients to manage their condition and treatment choices and use.
3. Provides insight into barriers and enablers for patients to move to the next logical step in their condition and therapy management—thus improving medication persistence and reducing time to reach optimal therapy.
4. Begins to provide insight into patient segments but often requires enhanced techniques for getting at the underlying attitudes and motivations and decision-making processes.

An example of a Patient's Journey is shown in Fig. 5.1.

Enhanced insight techniques can reveal faulty decision-making, incomplete information, or a lack of comprehension that can impact the adoption of the intended treatment, new devices or systems, or the patient's decisions to "stay the course" and complete the intended treatment plan.

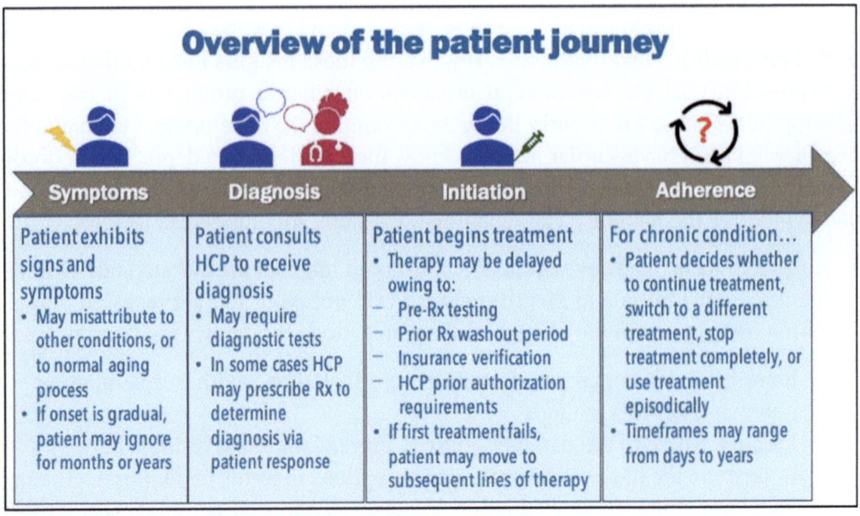

Fig. 5.1 Example of a patient's journey

- *Discovery phase*: Use tools such as reflective journeys, caregiver observations, ethnographic observation, and *codiscovery to uncover the often surprising and deep thought processes your prospective patient has about their condition and treatment options. Be sure to incorporate your conceptual solutions.*
- *Reflective journeys*: Ask patients to recall and document all aspects of their diagnosis, their treatment history, and the introduction of their current treatment. Have a trained moderator review their reflective journey and delve into what was going through their mind at each step—particularly around treatment introduction and adoption. Reflective journeys work very well in uncovering issues that cannot be identified in direct questioning, often because humans are not fully conscious of why they made *specific choices, or even conscious of making those choices.*

Codiscovery

A method of obtaining deeper insights from research subjects, in which subjects are shown video footage of themselves and asked to narrate what they are doing, why, what thoughts they had at the time of the behavior, and what motivated them to make specific choices. For example, a Crohn's Disease patient might be shown a clip of himself massaging the injection site before self-injecting, and asked how he got into that habit, how it contributes to his injection experience, and so on. This narrative track added to the video footage helps probe on behaviors patients may not even have been conscious of doing at the time, and helps them identify their reasons for doing so, to better uncover unmet needs.

- *Ideation phase*: The synthesis to identify insights is a separate process that requires the relevant team to fully internalize the collected information/data; workshop it or conduct generative sessions to distill the primary drivers of behavior/decision-making; and then articulate the insights.
- Insights are a written expression of a consumption truth. The truth offers a certain amount of information for understanding the consumer's situation involving a subject (the topic) and the motivation, which together comprise the consumer's need. An element of tension is then needed to transform the need. An insight conveys a dilemma around a given subject, written in human terms. (reference: Insight Academy, Paris, France, www.insightacademy.fr/index-en.html).
- Writing a compelling insight should:
  - State the situation: "We **enjoy** using our outdoor pool but are often **bothered** by mosquitos."
  - Articulate the frustration: "I dislike using **personal** repellants on my children, because I **fear and believe they are unsafe**."

  - State the future desire: "I wish I could safely and effectively **protect** my children from mosquitos **around the pool** so that they can **enjoy** the pool, and I can **worry less**." (Reference: *Thrive*, Jonathan Dalton, www.thrivethinking.com/our-people/jonathan-dalton).

- Insights incorporate beliefs, hypotheses or speculations, and meaningful facts. An insight is not designed to tell a consumer what they should do but instead to connect with consumers in a meaningful way such that they will agree with your presentation, because it meets their needs and is deeply understanding of them as humans.

- *Testing phase*: In this phase, the goal is to test various alternative solutions to close gaps.

  – Create prototype solutions—for example, conceptual statements, rough drawings, storyboards, or video to demonstrate. The ability of your prospective patient to fully understand the product offering and relevant instructions for use is paramount to their proficiency to evaluate it or provide insight for their reactions to it.
  – Share the prototypes with your patients/users and have them compare and contrast with their current behavior to reveal obstacles and opportunities for these potential solutions.

- *Refinement phase:* If time and budget allow, test refined potential solutions—ideally with User Experience (UX) techniques that simulate real-world conditions for product usage.
- *Evaluation phase*: post commercialization, measure Key Performance Indicators (KPIs) to determine whether users are employing solutions as intended, and whether gaps remain. Market research methodologies may be incorporated along with Real World Evidence (RWE) patient data, and data from incoming Medical Affairs queries.

In the next section, we will share how these processes were followed in some real-world case studies.

Case Study 1: Testosterone Treatment Delivery Improvement

This first case discusses insights uncovered while a delivery system was in early development. Discovering the insight identified a potential barrier to the adoption of the new delivery system.

Situation

First, consider that the path to identifying low testosterone as a condition or embracing the diagnosis is wrought with difficult emotions for a 40+ male. Many, due to the pervasiveness of Erectile Dysfunction treatment and the complexity of that issue, will assume that is the problem and try that solution. The other symptoms—loss of muscle and bone mass, extreme fatigue, reduced ability to concentrate—may not be as motivating or easy to solve as reduced libido and performance issues.

Once the patient has been diagnosed, and treatment offered and accepted, he is required to learn how to use the treatment and be conscious of its precautions. Unlike the discretion of an Erectile Dysfunction pill, testosterone treatment is public within a household due to the caution and waiting periods involved.

So, a sensitive issue and diagnosis is matched with a treatment that is not particularly discreet or private. In human behavior, this is a less-than-ideal combination for success.

- Male hormone replacement therapy is delivered via patch or a gel that is applied topically (shown: AndroGel®, from Ascend Therapeutics). The gel can be packaged in single-use packets or tubes, or in an underarm applicator or pump.
- Prior to the availability of the underarm applicator or the pump, the process of applying the treatment via packet was deemed messy and had the potential for not dispensing the full dose.
- Importantly, with the use of topical testosterone treatment, caution is needed to avoid transferring the treatment to others via skin-to-skin contact or not sufficiently rinsing away/disposing of excess. Both the underarm application and the pump were designed to create more self-enclosure of the gel and to accurately dispense proper dosages, and to relieve the messiness.

Approach

1. Ethnographic observation of current application modes and behavior was conducted and filmed. In addition, a codiscovery discussion with patients using treatment occurred as the observation was underway.
2. Prototype testing in two phases occurred. The prototypes were "looks-like" representations—rough plastic models without operational parts, but able to be held and handled. The first phase consisted of individual interviews that exposed the new administration prototypes to current treatment users and asked them to imagine use, storage, etc. The second phase took the most likely administration candidate prototypes back to the above-mentioned ethnographic participants for evaluation and assessment.

Findings

During prototype development, patient concerns arose that were not anticipated, including very strong concerns about transference, i.e., transferring medication to a loved one such as wife, child, or grandchild. Owing to the familiarity of pumps/underarm applicators for soap dispensing and lotion dispensing, many men feared that the pump would be accidentally confused with other more benign products. Packets were easily stored in private drawers in a bedroom rather than in a bathroom that would be shared. With packet use many men believed they could have more control over product disposal and their own hand washing and wait time. Hence, neither the underarm applicator nor pump were guaranteed to be an improvement for a portion of users. The fear/concern of misuse/mishandling outweighed the ease of application benefit.

Results

While the new administration options were introduced, the packets also remained on-market as a sizable segment of users preferred the original administration mode to a new and different one.

Case Study 2: Rare Disease: Is IV Administration a Real Problem?

This case demonstrates that preconceived concerns about intravenous (IV) administration are not always true and that the benefit of being in-home and ambulatory is worth some extra work for some patients.

Situation

- The treatment is the first specifically designed for this condition and demonstrated very high efficacy in Randomized Controlled Trials.
- The medication is a monoclonal antibody; owing to the orphan disease designation and fast-track filing, the intravenous delivery was best for Randomized Controlled Trials and quickest to market.
- This acute treatment requires a 60-min IV in an infusion location, with a waiting period afterward.
- The IV is dosed once every 3 weeks for 6 months.

- Because it is a large molecule biologic, self-administration/subcutaneous delivery was considered somewhat difficult to achieve from a technical perspective.
- While looking to attain a self-administered subcutaneous delivery option, the team investigated interim solutions: an HCP-administered subcutaneous vial, as well as on-body infusion pumps.
- Patients are visually impaired, and competition was looming with a subcutaneous option.

Background

IV therapy can be accomplished in a number of different ways: (a) where a healthcare professional injects via a syringe directly into a vein (referred to as a "push"), used when an immediate drug delivery is required; (b) via an infusion port administered more slowly, often to a patient sitting comfortably in a recliner chair (where slower administration is appropriate); or (c) via an on-body infusor—an adhesive (or belt)-attached device, which can allow slower, ambulatory infusion of a medication.

Approach

Ethnography was conducted with users of existing administration vehicles in other disease states—on-body infusers were used in Immunoglobulin-G4-related diseases and hemophilia—to understand how long it took patients to learn how to handle, and what they defined as the key benefits. Notably, these are chronic conditions, whereas our condition of interest is acute.

Illustrations, videos, and instruction diagrams were shared via focus group discussion with HCPs who treat the condition of interest.

Revised diagrams and instructional directions were included in demand surveys among patient prospects and treaters.

Findings

In evaluating the potential options for an alternative to the traditional IV, the traditional IV was not a big problem, because treatment was finite. Most patients preferred an in-home option, however, owing to the logistics of an infusion center visit. The HCP-administered option, despite a slow push delivery, was preferred if it could be done at the same time as the exam visit. Even though the on-body infuser required inserting a syringe and the treatment needed to be mixed and put into the syringe, it appealed to patients, because it was both in home and allowed the patient to be ambulatory while the treatment was dispensing.

Results

The on-body device moved into development to provide an in-home, ambulatory option to ward off the potential competitive threat of self-injection. Given the size of the molecule the company hypothesized that true self-injection would be very hard to achieve, at least in a short-term window.

Case Study 3: Are More Delivery System Options Better?

Sometimes, additional choices in devices do not necessarily add value—even if newer technology is available. While engineers may be keen to develop cutting-edge delivery systems, ultimately, it's patients who decide which devices solve an unmet need for them.

Situation

The manufacturer of a self-injected biologic medication was exploring the opportunity to bring a third delivery system to market. The product's existing delivery systems—a subcutaneous syringe, and an autoinjector pen that housed the syringe—had been on the market several years, but technological advances now provided the manufacturer with the opportunity to introduce new delivery options. Patients with certain conditions had been given a choice between the two existing devices for this product, while patients with other conditions had been offered only one device (the autoinjector). The device engineering team was investigating several promising technologies; however, they needed to narrow down their options, and requested commercial market research to help them understand which device attributes would be most useful and impactful in solving patients' injection needs.

Approach and Findings

The market research team proposed an in-home ethnographic study to observe patients currently using various subcutaneous self-injectable medications and devices. Objectives for the study were to understand:

- Which device types were most user-friendly for various patient types?
- What unmet needs existed across devices?
- Which aspects of the product peripheral to the actual administration—for example, product storage—affected ease of use?

This study yielded some surprising results and led to a quantitative (conjoint) analysis of all patient types, examining all potential device attributes.

Overall findings included:

A. Trypanophobia, or fear of medical procedures involving needles, affected a number of patients. For some of them, the autoinjector pen helped address the problem by keeping the needle hidden from view; others had family members or friends administer their injections for them. Potential device types that did not involve needles, or kept needles hidden from view, were more of interest to them.

B. Three types of pain were involved in self-injecting: the needle stick as it penetrated the skin, the acidity of the medication itself, and the pressure from the volume of medication accumulating underneath the skin.

   - Some patients seemed more sensitive than others to a burning sensation due to the pH of the medication, but no specific characteristics emerged determining who would or would not have this sensitivity.
   - Patients with higher body mass index (BMI) were less likely to experience the third type of pain, while those with lower BMI were likelier to feel the pressure of the bolus of medication.

C. The key trade-offs between syringe and autoinjector centered on control versus efficiency. The person likeliest to prefer the syringe had:

   - Sensitivity to the acidity and/or the bolus of medication and a desire to reduce the rate of flow by administering the injection more slowly, thus reducing the pain.
   - Indifference to needles/syringes.
   - A need for control over every aspect of the injection process, including rate of speed/pressure on the plunger, angle of syringe, ability to see the medication flowing out of the syringe, pressure on the injection site, ability to pause during the injection process, and so on.

   The person most likely to prefer the autoinjector was more inclined to simplicity and was happy to allow the device to determine all the above factors. In addition, these patients either were less sensitive to the pain aspects of the injection or wanted the fastest possible process to get the injection over with quickly. Some patients also exhibited a fear of seeing the bare needle of the syringe, which the autoinjector concealed within a sheath, and a few voiced a fear of making a mistake with the more complex syringe process.

D. The concept of "convenience" was multilayered, and far more complex than the manufacturer had realized. It included.

   - Frequency of dosing combined with ease of remembering to dose (thus, an injection every week was "more convenient" than an injection every other week, because it happened the same day of each week).

- Size of the device packaging, for storing in the refrigerator or carrying in a purse or briefcase to inject at the office.
- Labeling of the packaging—some patients disliked others in their home opening the fridge and seeing the name of the medication prominently visible.
- Ease of opening the package, opening the device, and opening the injection instructions.
- Whether the product was available for pickup at a retail pharmacy or required shipping from a specialty pharmacy—and consequently whether someone needed to be at home to sign for the medication and refrigerate it immediately.
- Ease of reordering.
- Portability of the medication, for patients who needed to travel with their treatment.
- Disposability of the used devices.[1]

E. Above all, what patients desired in delivery systems was choice. Being able to select from more than one delivery system gave them a sense of control over the administration of the medication and allowed them to prioritize their most critical needs. But having more than two device options did not increase patient affinity for the product or address significant unmet needs, and in some cases created confusion.

Results

The upshot from this exploration is that the manufacturer decided not to make the investment in additional drug delivery systems. They did pursue reducing the acidity level of the medication's formulation, plus invested in a sharper needle, to address two of the causes of injection site pain.

Case Study 4: Lab Switching from Custom to Off-the-Shelf Syringe

This case focuses on hospital staff device usage, but it illustrates some critical points that may be relevant to patient audiences as well.

---

[1] Market research was conducted in multiple geographies, in which local regulations for used needle disposal varied considerably. Some of the patients interviewed were unaware of or misinformed about those regulations and were not disposing of their used devices appropriately—yet another opportunity for the manufacturer to close a gap. It was noteworthy that patients in the Los Angeles area were especially concerned about illegal drug users rummaging through their garbage bins for used syringes—and less concerned about potential contamination within the household.

1. Assumptions about device changes need to be tested—ideally under real-world conditions—to understand user acceptance of the new device, changes needed to the process of using the device, and gaps in understanding that must be filled with appropriate communications and training.
2. When dealing with a combination product, in which the delivery system impacts the effects of the medication, and the medication's characteristics in turn impact the delivery system, it is imperative to consider their interactions holistically, rather than separately.

Situation

A diagnostic product used in hospital labs was manufactured as a prefilled syringe, which the technologist would inject into the patient's IV line before administering a test. Timing of the injection was critical, because if the medication was injected too quickly, the patient could experience an adverse event. To reduce this likelihood, the manufacturer had developed a customized syringe with some specific features to assist the tech in delivering the medication at a slow, even rate:

- Greater resistance due to a tighter fit of the plunger within the tube.
- Larger flanges, for a more secure grip and better leverage.
- A slightly smaller volume in the tube, resulting in two key considerations:
  - The tube was completely filled with the medication, leaving no room for air.
  - The pressure in the tube required a quick, short pull on the plunger to prime the syringe before injecting.

The cost of producing the custom syringe was significant enough that the manufacturer decided to explore switching to an off-the shelf syringe, which would save considerable expense over the remaining life of the product's patent. Commercial market research was requested to study the potential impact on resulting usage and brand preference from the switch.

Approach

The market research team determined that a handling study would be most informative in identifying how technologists would respond to the new syringe. To simulate conditions in the hospital lab, the team assembled a mannequin with an IV unit, and provided the participating technologists with one of the commercially available custom syringes, as well as two of the proposed generic syringes; all three were filled with the actual medication. Participants were asked to administer an injection into the IV line with the current syringe, then inject with the proposed syringe, and finally inject with the proposed syringe while narrating aloud the steps they were taking and their reactions.

Findings

The technologists were able to use the proposed generic syringe without too much difficulty, and nearly all stated they would be able to get used to the new syringe after a week or two of usage (longer for labs with smaller volumes). Of note, however, were a number of differences in the syringes that affected the injection process and could have presented significant issues if the technologists were not properly informed and trained on the new device. These differences included:

- A slightly larger volume in the tube, leaving room for air, which—for patient safety—would need to be removed before initiating the injection.
- Less resistance on the plunger, requiring the technologist to push the plunger more carefully to avoid injecting too quickly and causing an adverse event.
- Smaller flanges on the end of the tube, which gave the technologist less leverage; depending on how the user preferred to hold the syringe during the push, this difference might or might not be noticed.

Participants strongly recommended both repeated communication with hospital lab staff prior to the switch, as well as in-person training of the lab staff to point out the differences in the syringe and the changes to the process.

Results

When the manufacturer calculated the cost of multiple waves of communication and in-person training needed at all hospital labs across the USA, it nearly equaled the potential savings of switching to the generic syringe. The deciding factor was the increased potential for adverse events with the new syringe. While not significant, it was nevertheless a risk to patients that the team considered undesirable.

Common Learnings This section is intended to share some truths that have surfaced after significant learning about prospective patients across multiple therapy areas and in various situations of adopting delivery systems.

Top Ten

1. Just because you think a new device is easier, better, more convenient, or more technologically advanced, your users may not want it. Ask them!
2. Workarounds are often easier to adopt and maintain than switching to improved delivery/devices.
3. Chronic versus acute/short-term use of treatment is a priority consideration. Patients in acute situations may see burdens or inconveniences differently from those with chronic conditions.
4. Even if users agree that a new delivery system has improvements over their current approach, they may not adopt it, because the change is too difficult and does not sufficiently resolve their unmet needs.

5. Consider who beyond the user may be affected—for example, family members, caregivers, colleagues—and incorporate their perspectives into the research. If possible, conduct research onsite (where the product is used) and obtain feedback interactively among all affected individuals.
6. Introducing a new process or device often requires System 2 thinking and can create frustration among users, but once they use it regularly for a period of time, it can become second nature (System 1 thinking).
7. Asking users what they do may not tell you what you need to know, because they may not be conscious of what they do (System 1). Observing them use a product in their own environment is generally the most thorough, detailed, and accurate way to learn how they actually use it.
8. For additional insight, show users footage of them using the product and have them narrate their thought process for each step; in some cases, they will realize steps that they perform unconsciously (System 1).
9. Communication and training on new devices are essential to their adoption. It is impossible to overcommunicate—particularly when giving users an advance warning before the switch.
10. Health literacy applies here, so communicate at the level of a sixth grader in order to assure understanding.

Finally, some resources for improved approach to incorporating patient perspective:

- Patient Journey:
  - Identifies key insights about how and why patients manage their conditions in the current market—allows you to see if there is an open opportunity.
  - Can zero in on drivers and barriers at each key decision point.
  - Also identifies sources of information used, and gaps in understanding at each stage.
- Existing patient segmentations, to understand the specific needs of distinct patient populations.
- Prototype assessments:
  - Observational, preferably in real-world environments.
  - Range in scope from low to high fidelity; that is, you do not have to have all the answers/bells and whistles to gain insights about interest, usability, and willingness to adopt.
- Basic qualitative exploration:
  - Explain the product to the user in terms of benefits, instructions for use.
  - At the end of the session, have the user explain the product as if to a friend or colleague.
- Demand/Life Cycle Management quantitative assessment to predict adoption share, weigh options, and monetize the solution.

Chapter 6
Expectations from Regulators

Shannon Hoste and Natalie Abts

Abstract The regulatory landscape plays a critical role in meeting patient safety requirements and developing the right products for purpose. Each global authority dictates their processes for the clearance/approval of medical products and the continued surveillance of these products in regions where they are marketed. It is important to understand what activities are required to support a regulatory package to demonstrate safe and effective use. The goal of this chapter is to explore considerations for maintaining user centricity within the regulated industries of medical devices, pharmaceuticals, and biologics, discussed from the regulatory purview and perspective. This includes understanding the role of a regulatory body, how regulatory science is used to address patient safety, considerations for human factors/usability engineering in different regions, and how developers can apply a robust, patient-centric process while ensuring timely delivery of products to market.

Keywords Use error · Hazard-related use scenario · Use-related risk analysis (URRA) · Use-related hazard

Introduction

Earlier chapters have explored the patient ecosystem, user needs, user centricity in design, business constraints, and commercial considerations. The regulatory landscape also plays a critical role in meeting patient safety requirements and developing the right products for purpose. Each global authority dictates their processes for the clearance/approval of medical products and the continued surveillance of these

S. Hoste (✉)
Pathway for Patient Health, Kymanox Corporation, Cincinnati, OH, USA
e-mail: Shannon.Hoste@Pathway4PH.org

N. Abts (✉)
Genentech, South San Francisco, CA, USA
e-mail: abts.natalie@gene.com

R. R. Nesbitt et al. (eds.), *Patient-focused Development of Biopharmaceutical Products*, AAPS Advances in the Pharmaceutical Sciences Series 27,
https://doi.org/10.1007/978-3-031-98253-8_6

products in regions where they are marketed. What activities discussed in earlier chapters, though, are required to support a regulatory package to demonstrate safety and effective use?

The goal of this chapter is to explore considerations for maintaining user centricity within the regulated industries of medical devices, pharmaceuticals, and biologics from the framework of the regulatory purview and perspective. When considering this, questions such as the following arise:

- What is the job of a regulatory body?
- What is regulatory science and how does it address patient safety risk and medical product viability?
- How is human factors/usability engineering considered across the regulatory landscape in different regions?
- How can a robust, patient-centered development process be balanced with competing goals of timely delivery of a product to market?

Regulator Purpose

What is the job of a regulatory body and of the regulators, reviewers, and inspectors that are engaged across a product's lifecycle? To explore this, let's look at the tasks of some of these organizations. Table 6.1 captures the mission statements for the United States (US) Food and Drug Administration (FDA), European Union (EU) European Medical Agency (EMA), and People's Republic of China (China) National Medical Products Administration (NMPA). Common threads emerge between these three regulatory authorities, namely, the need to ensure:

- The safety, efficacy, and security of drugs, biological products, and medical devices, including market authorizations and ongoing surveillance.
- Innovations that make medical products more effective, safer, and more affordable.
- Accurate, science-based information for drugs, biological products, and medical devices.

In reviewing these mission statements, it also becomes clear that the goals of the regulators are patient centric, with clear in priorities around ensuring that their citizens have:

- Access to medical products that are consistently safe and effective.
- Timely access to new innovations that improve the standard of care.
- Product information that is accessible, accurate, and understandable.
- Products that support a resilient health-care system.

Understanding this perspective and the driving factors for regulators can help developers to focus on the most critical inputs for addressing patient safety and identifying the right data to answer the most common regulatory questions.

Table 6.1 Exploration of regulatory authority mission statements

US FDA[a]	The Food and Drug Administration is responsible for protecting the public health by ensuring the safety, efficacy, and security of human and veterinary drugs, biological products, and medical devices, and by ensuring the safety of our nation's food supply, cosmetics, and products that emit radiation. FDA also has responsibility for regulating the manufacturing, marketing, and distribution of tobacco products to protect the public health and to reduce tobacco use by minors. FDA is responsible for advancing the public health by helping to speed innovations that make medical products more effective, safer, and more affordable and by helping the public get the accurate, science-based information they need to use medical products and foods to maintain and improve their health. FDA also plays a significant role in the Nation's counterterrorism capability. FDA fulfills this responsibility by ensuring the security of the food supply and by fostering development of medical products to respond to deliberate and naturally emerging public health threats.
EU EMA[b]	The mission of the European Medicines Agency (EMA) is to foster scientific excellence in the evaluation and supervision of medicines, for the benefit of public and animal health in the European Union (EU) Facilitate development and access to medicines. Evaluate applications for marketing authorization. Monitor the safety of medicines across their lifecycle. Provide reliable information on human and veterinary medicines in lay language.

(continued)

Table 6.1 (continued)

 China NMPA[c]	(1) To supervise the safety of drugs (including traditional Chinese medicines (TCMs) and ethno-medicines, the same below), medical devices and cosmetics; to draw up regulatory policy plans, organize the drafting of laws and regulations, formulate normative documents, and supervise the implementation thereof; to research and formulate regulatory and supportive policies that encourage new technologies and new products for drugs, medical devices, and cosmetics. (2) To undertake standards management for drugs, medical devices, and cosmetics; to organize the formulation and publication of the Chinese Pharmacopoeia and other drug and medical device standards, organize the drafting of cosmetic standards, organize the formulation of the classification management system, and supervise the implementation thereof; to participate in formulating the National Essential Medicine List, and assist in the implementation of the national essential medicine system. (3) To regulate the registration of drugs, medical devices, and cosmetics; to develop the registration system, conduct strict review and approval for marketing, improve measures to facilitate the review and approval process, and organize the implementation thereof. (4) To undertake quality management for drugs, medical devices, and cosmetics; to develop Good Laboratory Practices (GLPs) and Good Clinical Practices (GCP) and supervise the implementation thereof; to develop Good Manufacturing Practices (GMPs) and supervise the implementation thereof in line with NMPA's responsibilities; to develop good practices on the distribution and use of medical products and guide the implementation thereof. (5) To undertake postmarket risk management for drugs, medical devices, and cosmetics; to organize the monitoring, evaluation, and handling of adverse drug reactions, medical device adverse events, and cosmetic adverse reactions; to undertake emergency response management for drugs, medical devices, and cosmetics in accordance with law. (6) To undertake management of qualifications for licensed pharmacists; to formulate regulations of qualifications for licensed pharmacists, and guide and supervise the registration of licensed pharmacists. (7) To organize and guide the supervision and inspection of drugs, medical devices, and cosmetics; to develop the inspection system, investigate and punish illegal activities during the registration process for drugs, medical devices, and cosmetics in accordance with law, and organize and guide the investigation and punishment of illegal activities during the manufacturing process in line with NMPA'S responsibilities. (8) To engage in international exchange and cooperation in the regulation of drugs, medical devices, and cosmetics, and participate in developing relevant international regulatory rules and standards. (9) To guide the work of drug regulatory departments of all provinces, autonomous regions, and municipalities directly under the Central Government. (10) To complete other tasks assigned by the CPC Central Committee and the State Council.

[a]https://www.fda.gov/about-fda/what-we-do#:~:text=FDA%20Basics-,FDA%20Mission,and%20products%20that%20emit%20radiation. Pulled 10/31/23
[b]https://www.ema.europa.eu/en/about-us/what-we-do pulled 10/31/2023
[c]http://english.nmpa.gov.cn/2019-07/18/c_377587.htm pulled 10/31/2023

Applying Regulatory Science

The FDA defines “regulatory science” as the science of developing new tools, standards, and approaches to assess the safety, efficacy, quality, and performance of regulated products. The regulatory science around assessment of drugs, biologics, and devices is established, but continues to evolve with new technological advancements. How, then, do developers determine the practical applications to meet these goals? Additionally, the challenge in the combination product space is the interaction of these varying regulatory requirements. How might independent device constituent parts affect the safety, efficacy, quality, and performance of the other constituent part(s)? How do you demonstrate, when combined into one product, that all requirements for each part of the whole have been met? Further, to what extent can developers leverage existing data to get products to market faster?

Development Decisions

The answers to these questions can be incredibly complex but are best addressed, and most valuable, when considered at the beginning of development. Previous chapters have discussed trade-offs in decision-making from various lenses, including the impact of new presentations, increased dosing complexity, and migration to home use. Consequently, these decisions also require regulatory trade-offs. The trade-offs from the standpoint of technical rigor are dynamic and wide-ranging, and the robustness of data needed to demonstrate product safety has great impacts, including the time to market and costs associated with development. Developers can think ahead and be more prepared to make these decisions by beginning with the end in mind.

To illustrate this point, imagine a medication is introduced to the market through IV administration. This is often one of the first drug delivery methods used for initial clinical trials and commercialized products. Since IV drug delivery utilizes existing and established medical practice, it usually does not necessitate the inclusion of more novel devices with the drug product. In this scenario, the regulatory focus is primarily the drug, and user interface considerations (e.g., packaging and labeling design) are minimal. However, things can become more complex when unique risks are identified, and more than the standard risk mitigation practices are necessary. Examples of unique use-related risk could include potential mix-ups between similar medications, complex dosing schemes, or novel administration techniques. These risks can trigger health authority concerns, and evidence may be required to demonstrate risk reduction. In these examples, standard clinical data supports safety and effectiveness when used correctly; however, different tools are necessary to assess that the product supports correct use. In such cases, regulatory authorities may review data from a human factors program utilized to demonstrate effectiveness of risk controls that address potential for medication error due to differentiation, dosing, and administration.

Expanding use of a medication outside of IV administration brings on more regulatory scrutiny as complexity is added. A common next step, if appropriate, is to incorporate a medical device for in-clinic use by health-care providers and/or use at home by lay people. Once a drug introduces new use-related risks, such as when it becomes a part of a combination product and/or when new users or use scenarios are introduced, a formal human factors program is required to assess use safety and provide evidence of risk minimization through a design-controlled process. Though more robust requirements increase time to market and may be perceived as putting additional burden on developers, introducing a combination product can greatly increase the scope of use and provide more desirable options for patients.

Developers may also identify benefits to migrating a medical device from exclusively in-clinic use to home use by patients. Potential drivers could include facilitating medication administration for patients during times of restricted travel, addressing clinical overcapacity, or reducing the scheduling burden for both patients and clinicians. In this scenario, even more attention is required to ensure that user needs are met, and that use-related risk is mitigated to the greatest extent to accommodate the new users, uses, and use environments. Decisions about the method of drug delivery, how the product is packaged, whether a device should be incorporated, and who will operate the device all impact the regulatory expectations and have implications for the level of rigor needed to demonstrate that safe and effective use has been designed into the product. To ensure success of this product, beginning with the end in mind, the planning will need to account for the methodology to demonstrate that this new product supports safe and effective use.

Human Factors and Usability Standards and Regulatory Framework

To address safety and effectiveness from a use perspective, human factors/usability engineering (terms are often used interchangeably) provides a framework for product ideation, development, and validation. Though the science of human factors does not focus only on risk, risk management drives meeting the regulatory and safety requirements related to product use. Developers can use a variety of tools to identify, analyze, and track decisions and trade-offs. Not every tool is appropriate for every product, though, and efforts should scale up or scale down based on product complexity and anticipated risk.

Human factors expectations are described globally via various guidance and standards. The U.S. FDA has published guidance documents and expectations for both medical devices and combination products, and also recognizes AAMI ANSI IEC 62366–1:2015 + AMD1:2020 (reference). Another approach to use-related safety is invoked within EU MDR. Within these European regulations, general safety and performance requirements primarily around use-related risk are set out in Annex 1 (reference Annex 1 General Safety and Performance Requirements, Chapter 1 General Requirements, Section 5):

> In eliminating or reducing risks related to use error, the manufacturer shall (a) reduce as far as possible the risks related to the ergonomic features of the device and the environment in which the device is intended to be used (design for patient safety), and (b) give consideration to the technical knowledge, experience, education, training, and use environment, where applicable, and the medical and physical conditions of intended users (design for lay, professional, disabled, or other users).

The goal of this chapter is not to reiterate the specific processes discussed in the FDA guidance documents and international standard, but to discuss key topics that should be considered by sponsors throughout development as they aim toward a regulatory submission, including regulatory pathway, the value of early-stage research, approaches to user testing, and the role of user interface components such as labeling from a regulatory and risk-reduction perspective.

Human Factors Development Process

Any product development process should include defining user needs, identifying solutions that could address those needs, and iteratively evaluating the user interface. Finally, objective evidence must be provided to demonstrate that this result has been achieved. Human factors data provides evidence that the user interface supports safe and effective use for the intended uses, users and use environments, and demonstrates that use-related safety has been built into the product. A graphical overview of a typical human factors process is captured in Fig. 6.1.

Key steps in this process include.

- *Use Specification*: Identify users, uses, use environments.
- *Use-Related Hazards*: Identify user interface characteristics related to safety and potential use errors, identify known or foreseeable hazards and hazardous situations, identify and describe hazard-related use scenarios.
- *Formative Assessments*: Evaluate user interface designs and identify potential use errors.
- *Summative (Validation) Assessments*: Evaluate final user interface, focusing on identified hazard-related use scenarios and safety-critical tasks.
- *Analysis and Documentation*: Update Risk Management File and benefit-risk evaluation, compile Usability Engineering File and HFE/UE reports.
- *Monitor Product Performance*: Proactive surveillance of product on the market, identify any changes to purpose or context of use, any new risks, etc.
- *URRA*: Risk-based tool captures and tracks user interface design and design decisions as they relate to use-related risk.

It is important for developers to think critically about their product needs when determining the scale and scope of activities that should be completed for each step

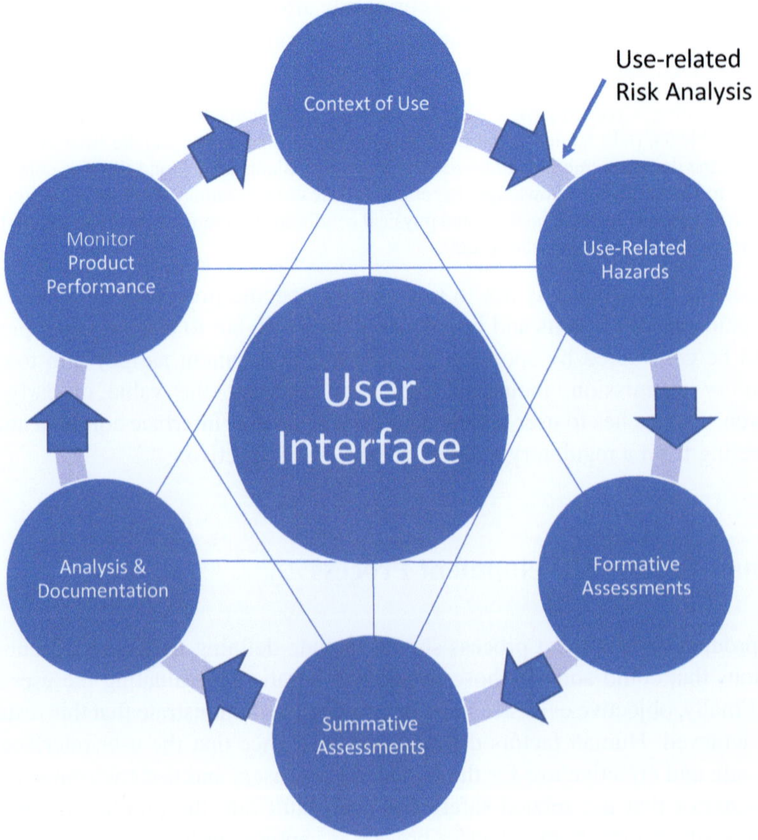

Fig. 6.1 Summary of the human factors process

in the process. Though they may be scoped differently based on product needs, the scientific tools and methods used throughout the process are consistent whether applying them to a drug product, combination product, medical device, Software as a Medical Device (SaMD), or other product type. What changes is the regulatory pathway, and subsequently, the approach required to meet the expectations of regulatory review. Table 6.2 includes examples of how approaches may differ based on regulatory needs and product differences.

As demonstrated above, the assessment of use-related risks is fundamental to any product development effort. Proper identification and assessment of risk require an understanding of users, use environments, and use contexts. Another important point to note is the expectations for formative assessments. Although in most cases, it is appropriate to conduct formative assessments, they are not typically part of the regulatory requirements as long as successful validation can be achieved. However, not only is it difficult to conduct a successful validation without having done formative work, it is also unlikely that the product will be optimized if the formative process is lacking. Even if the regulatory bar is met for minimizing risk and

Table 6.2 Examples of regulatory pathways (FDA focused)

Product type	Objective for regulatory submission	Human factors approach
New drug approval, with device constituent part (e.g., autoinjector, inhaler, nasal spray)—Unique users, use environments, uses, or user interface	Demonstrate safe and effective use	URRA: Required Formative studies: Highly recommended, but not required HF validation study: Required to study interaction of unique product characteristics
New Drug Approval, with device constituent part—minimal differences with approved product	Demonstrate safe and effective use	URRA: Required Formative studies: May not be needed HF validation study: Maybe; comparative data in the form of a threshold analysis may be substituted for HF validation
New Drug Approval, with device constituent part—modified user interface	Demonstrate safe and effective use of modified components	URRA: Required Formative studies: Recommended for modified components of user interface HF validation study: Not required if differences are minor and not safety related; if needed, a modified HF validation may be conducted to study only the user interface changes
Generic drug with device constituent part	Demonstrate sameness with reference drug	URRA: Required Formative studies: May not be needed HF validation study: Maybe; comparative data in the form of a threshold analysis may be substituted for HF validation
New drug approval, with device constituent part utilizing AI/ML[a] (e.g., Physiologic Closed Loop Control (PCLC) System)	Demonstrate safe and effective use	URRA: Required; must consider specific risks for PCLC systems, such as complacency and loss of situational awareness Formative studies: Highly recommended, but not required HF validation study: Required if comparative data cannot be leveraged; must cover specific regulatory questions around safe use of AI/ML identified through guidance and presubmission interactions
New drug/biologic approval or device clearance/approval with an SaMD	Demonstrate safe and effective use with the SaMD constituent part	URRA: Required; must consider specific risks associated with SaMD operation; operating platform considerations must also be included in risk assessment (e.g., alerts from a continuous glucose-monitoring system received on a patient's mobile phone) Formative studies: Highly recommended, but not required HF validation study: Required if comparative data cannot be leveraged; must cover specific regulatory questions around safe use of SaMD identified through guidance and presubmission interactions

[a]Artificial intelligence/machine learning

developing a product that is safe enough for approval, that does not mean that the product will be easy or satisfying to use, or that it will perform as well as expected on the market. For example, there may be low-risk tasks with navigating software components of a user interface, but that does not guarantee that menus will be easy to navigate, leading many users to complete tasks through trial and error and causing delay. Thus, developers should carefully consider the scope of formative work based on both user needs and regulatory requirements so that these critical activities can be incorporated into the development plan.

Early-Stage Research

Research completed at early stages of development (also referred to as "discovery research" in earlier chapters) is a good example of the type of activity that may not be strictly required by regulatory bodies but provides immense value and can have a substantial impact on creating the best product for purpose. Examining the details of the user/patient journey not only helps us to understand what considerations might be necessary for patient safety, but also helps us to explore non-safety-related aspects of device use such as environmental factors that may impact correct use and satisfaction, translating to better-designed products.

A vignette from Chap. 2 illustrates this point. Mary's struggle with issues such as waiting for an appointment, limited face-to-face time with a physician, and insurance constraints are real issues that encompass the patient's journey and are critical to understand the full extent of the patient experience. However, understanding and accounting for these factors are not necessarily required to achieve regulatory success, and safe and effective products can be produced without fully understanding the expanse of the systems in which products function. The pen injector Mary will use to administer her medication must, at minimum, go through successful validation testing and demonstrate risk minimization. This includes all elements of the user interface, such as any instructions for use (IFU) and quick reference materials. However, it is not required to demonstrate that the user research that may have yielded detailed information about Mary's experience has been performed, even though this understanding may lead to designs that increase patient satisfaction and potentially even medication compliance.

Another example is a home-use injectable that might be presented in a large carton requiring refrigeration. Through ethnographic work, one might find that home users often struggle to find a storage place in the fridge, especially when multiple devices are obtained at once. This could lead users to keep products outside of the required temperature range, or remove devices from carton for storage, potentially exposing the medication to light. While a large carton is unlikely to be the sole barrier to regulatory approval, and these are cascading risks that some might consider out of the manufacturer's control, these design decisions still impact the lives of patients. Failure to consider these factors may affect user acceptance, medication adherence, and postmarket complaints.

Importantly, the regulatory focus on safety for human factors evaluations does not imply that early-stage research, or anything outside of the scope of safety, is not of a concern to health authorities. Sponsors are required to characterize the users, use environments, and other aspects of device use for products to reach market. As gatekeepers for keeping unsafe products and medicines out of the hands of consumers, though, nonsafety concerns become more of a marketing risk for developers than a barrier to approval. Thus, the onus is on developers to be accountable for going beyond the regulatory expectations and practice more robust and holistic processes to develop the best products and resources for users.

Formative Approaches

The outcomes of early-stage research can feed into selection of off-the-shelf device components, or the decision to develop a new product. In either case, sponsors will need to consider the magnitude of human factors evaluations required, first in the formative stages, and eventually for product validation.

Of the human factors process steps, the formative assessment stage will likely have the most variation in scope depending on the product. This applies not only to the number of formative studies or activities executed, but also to the methods in which they are achieved. Activities for newly developed hardware or software may start before prototypes are even created, utilizing concepts or storyboards to obtain feedback. Once prototypes are available for user interaction, formative studies are designed based on the goals appropriate for the stage of development and level of fidelity. The participants in each activity should also be customized as appropriate. For example, expert reviews by human factors specialists (i.e., heuristic evaluations) are a low-cost activity that can be done quickly to identify potential design improvements before putting the product in front of the end users. More traditional formative studies in which participants interact with the product in a realistic manner can also be customized in many ways to meet product needs. Table 6.3 describes examples of formative study methods.

Gathering formative data has a substantial impact on development, and selecting the right activity and scope can make a big difference in both safety and usability of the final product. However, health authorities do not typically have specific requirements for formative work. In theory, a product could get to market successfully as long as an URRA is completed, and a validation study produces acceptable results. Additionally, for well-understood devices, data from previous human factors studies can be leveraged, and executing additional formative studies may have little value if differences are minor. For new developments and novel situations, though, it is unlikely that a validation study will be successful if formative assessments are limited or omitted. Sponsors must make their own decisions on whether formative assessments are needed, and which methods will produce data to fill any gaps. Though specific formative activities are not required in every situation, many of the standards and guidance documents do reference these activities, and health authorities such as the FDA may expect to see details about the formative process and how

Table 6.3 Formative study methods

Study method	Description	Advantages	Disadvantages
Think-aloud protocols	Study how participants describe what they are thinking while they perform tasks with the product	Immediate feedback on product use that avoids the potential impact of memory decay	Initial feedback at the point of use may not be well-thought-out, and users might change their perspective after further use
Labeling comprehension	Focuses on assessment of product labeling (e.g., IFU, device label, carton labels) typically using knowledge-based questions	Focus on labeling alone removes confounding variables	Participants have less context around device use Searching for certain information in the labeling may not be reflective of real-world use
IFU mandatory	Participants are required to use the instructional materials for each task step	Allows for focus on the labeling while providing more context for the user than knowledge-based questions alone	Does not assess the most likely real-world use scenario Some users struggle to use the IFU for every step
Comparison studies[a]	Evaluate multiple prototype options with the same or different users	Helpful to narrow down device selection between multiple options, and to obtain user perspective on design trade-offs	More users are required for *"between-subjects"* studies Easy to introduce bias during *"within-subjects"* studies
Team studies	Recruit different roles who may work together in the same use environment and complete tasks as a group	More realistic simulated use Ability to identify issues that may not be detected by isolating roles	Scheduling and recruiting difficulties Unrealistic interactions if team members do not work together in the real world

[a]This term is not referring to specific comparative use human factors studies as described in the FDA draft guidance *Comparative Analyses and Related Comparative Use Human Factors Studies for a Drug-Device Combination Product Submitted in an ANDA: Draft Guidance for Industry*

the data impacted product design (https://www.fda.gov/regulatory-information/search-fda-guidance-documents/applying-human-factors-and-usability-engineering-medical-devices).

Human Factors Validation

Both early-stage research and formative human factors assessments provide incredible value in ensuring a product is optimized for use and that risk has been mitigated to the extent reasonable or possible. Once a developer believes that it has been achieved, a human factors validation study is used to confirm the final product will support safe and effective use These studies, described in detail in the 2016 human factors guidance (https://www.fda.gov/regulatory-information/search-fda-guidance-documents/applying-human-factors-and-usability-engineering-medical-devices), usually involve simulated use evaluation of the final user interface by the intended users in a representative intended use environment, and include evaluation of unassisted use scenarios that encompass all safety-critical user tasks. This full-scale approach is expected not only for new products coming to market, but also marketed products that have substantial differences to the device end users, environments, or user interface.

Because the FDA and other regulatory bodies examine products holistically rather than piecemeal, it is the interaction of unique inputs to device use that must be considered when identifying the regulatory expectations to meet patient safety requirements. For example, a sponsor might believe that a prefilled syringe already on the market for home use would also be considered validated for a new indication, provided that the user interface is the same. The sponsor might also believe that they have already validated the capabilities of the user population for the new indication if they have other products marketed to that user group. They might also consider the labeling and IFU for the new product to be validated if they have already been assessed by other lay users. However, if the specific device users, use environments, and user interface have not been validated together, regulators may see a gap. Some factors that might trigger the need for a full-scale human factors validation include:

- New graphics and language used in the IFU.
- Anticipated differences in experience with similar devices for the intended user population.
- Modification of task order.
- Higher injection forces.
- Use in an emergency environment or scenario.

Although the regulatory bar can be high if seeking to avoid a full-scale validation of the entire user interface to demonstrate optimized patient safety, there are some opportunities to leverage depending on your product and regulatory pathway. One such instance is when modifications are made to an already approved product, and thus a modified device evaluation is appropriate. In these cases, only specific aspects of the user interface are changing, so only the data between the legacy product and new product requires validation. The unchanging aspects have already been deemed safe and effective by regulators, and thus the risk remains the same, assuming that new uses, user groups, or use environments are not introduced.

For modified device validation, an abbreviated version of a standard human factors validation study can be conducted, which assesses only the changed components. There may be some unique considerations, however, in comparing product versions. This is particularly applicable for cases in which modifications were made to address safety issues identified with the on-market product. Regulators are interested in the impact of the change on the end users and their perspective on how the differences might impact safety.

When it comes to leveraging human factors data, the theme of trade-offs once again arises. Developers must make decisions that could impact time to launch but may have substantial benefits to patients. For example, making no modifications to the user interface may increase the probability that a new validation study will not be required, but the sponsor loses the opportunity to make potential improvements. While getting devices in the hands of patients as quickly as possible is often imperative, sponsors must consider what they might be sacrificing. Opportunities to make incremental changes could ultimately make a big difference for patients, whether or not they are safety related.

Product User Interface

While it is important to know details of the human factors process and select the correct methods to develop and evaluate a product, it is also important to understand what falls into the scope of assessment and of regulatory purview. The physical device or combination product components, or the software interface if developing SaMD, are main components that users will interact with to achieve their end goal. However, the user interface encompasses more than that. FDA defines the user interface as follows (https://www.fda.gov/regulatory-information/search-fda-guidance-documents/applying-human-factors-and-usability-engineering-medical-devices):

- User interface: All points of interaction between the user and the device, including all elements of the device with which the user interacts (i.e., those parts of the device that users see, hear, touch). All sources of information transmitted by the device (including packaging, labeling), training and all physical controls and display elements (including alarms and the logic of operation of each device component and of the user interface system as a whole).

Because the entire user interface must be assessed during human factors validation, sponsors should consider all of these components during the early stages of development. If user interface components are assessed for the first time during validation, it is likely that unexpected problems could arise. Many of the formative activities described above can be applied and customized to different user interface components, whether evaluated independently or separately. However, the regulatory expectation is that all components will be evaluated together during validation.

It should also be noted that the user interface does not include every resource that a patient or user may have at their disposal. If we think back to Mary's experience on her diagnosis and treatment journey, she encounters resources such as a product website, which she accesses to obtain resources to learn more about her medication. Similar resources such as promotional materials, videos posted on sponsor website, and product training that a user could receive from a nurse or pharmacist but cannot be guaranteed, fall outside of the user interface scope, and are not typically required as part of formal human factors validation. Only materials that are packaged or contained directly with the product are considered part of the user interface, and other resources are not subject to the same rigor of assessment. However, developers should still carefully consider that design of any external resources and take a patient-centered approach by applying the appropriate human factors lens.

Labeling Components of the User Interface

Product labeling is a critical component of the user interface for any medicinal or device product, regardless of the regulatory pathway. In this chapter, we will be using the U.S. regulatory term for labeling. While the term "label" can be used in a more specific sense to refer to printed, written, or graphical material on the immediate container, the term "labeling" is used more broadly to also encompasses on-device labels and any additional written materials accompanying the product. Common components of the labeling include the IFU, quick reference guides, package inserts, and prescribing information. As end users will interact with some or all of the labeling before, during, or after product use, the design of these components can have substantial impact. As with the design of the physical device, the regulatory focus is on the aspects of labeling that could lead to patient or user harm if design is not optimized.

Though labeling is an important resource for patients and a critical tool to promote correct use, a common mistake is to view labeling as the main source of risk mitigation. The FDA, by way of ANSI/AAMI/ISO 14971, views it as a tertiary mitigation strategy (Table 6.4).

Regardless of the stringency of a specific market's regulations, this strategy is recommended to ensure that the strongest risk mitigations are implemented whenever possible. It is important to note that even though labeling design should not be the primary mode of risk mitigation, it is a critical tool when use errors cannot be prevented or addressed through design and can also provide the user with more opportunities to perform tasks correctly if design-based risk mitigations applied to

Table 6.4 Risk management prioritization and examples

1. Inherent safety by design	2. Protective measures in the medical device itself or in the manufacturing process	3. Information for safety
Use specific connectors that cannot be connected to the wrong component. Remove features that can be mistakenly selected or eliminate an interaction when it could lead to use error. Improve the detectability or readability of controls, labels, and displays. Automate device functions that are prone to use error when users perform the task manually.	Incorporate safety mechanisms such as physical safety guards, shielded elements, or software or hardware interlocks. Include warning screens to advise the user of essential conditions that should exist prior to proceeding with device use, such as specific data entry. Use alerts for hazardous conditions, such as a "low battery" alert when an unexpected loss of the device's operation could cause harm or death. Use device technologies that require less maintenance or are "maintenance free."	Provide written information, such as warning or caution statements in the user manual that highlight and clearly discuss the use-related hazard. Train users to avoid the use error.

the physical device fail. Additionally, all labeling is not created equally, and different aspects of labeling can be used strategically to mitigate most effectively.

An example of this is the use of on-device labeling. This strategy can be a more impactful source of risk mitigation than written materials placed inside the packaging. If users do not choose to use the instructions or other materials, the on-device label can catch their attention. However, many combination products have limited real estate on the physical device, and the need for other required labeling components such as bar codes and expiration dates often leave no space for additional information.

The next step could be to examine the product carton for opportunities. Smaller cartons may also have limited options, but a simple cue like a warning, caution, or note may be possible. The underside of the carton flap could also be utilized if it is large enough to accommodate printing and be salient to the user. Information like special dosing instructions or a quick reference guide provides the user with the opportunity to see those cues before they begin interacting with the product and can be helpful for users who may not normally open the instructions for use.

One must also consider that when users encounter the IFU, many may still choose not to open it. This could be for a variety of reasons, which could include believing they understand the product and do not need further instruction, or a desire to figure out product use on their own. However, there are also design-based reasons that could influence this choice, and sponsors can use strategies to increase the likelihood that a user might reference them. Considerations include:

- Location of the IFU in the package.
- Information printed on the unopened IFU.
- Ease of opening and unfolding.
- Compatibility with use environment (e.g., clean room for pharmacy-prepared products).

By putting critical thought into labeling solutions and removing barriers to use, developers can demonstrate their emphasis on risk minimization to regulatory bodies. It should be noted, however, that adding more mitigations and duplicating information in multiple places is not always a good strategy and may not meet regulatory needs. Sponsors should carefully consider trade-offs when determining the best risk mitigation strategy. For example, adding more text to a carton, if even safety related, may draw the user's attention away from other important information. Keeping this in mind, developers must explain the selected risk mitigation strategies thoroughly in their regulatory submissions so that health authorities can understand the full story of why each mitigation is considered effective, and why other solutions may have been ruled out.

Because the design of the physical device's constituent part(s) and the labeling components of the user interface are both critical considerations for developers, both should be considered at each stage of the HF process as appropriate. Table 6.5 explores some ways in which this can be done to achieve an optimized user interface that both meets regulatory expectations and meets user needs beyond any safety implications. This is not a comprehensive list of considerations but includes some of those that might be commonly encountered.

Table 6.5 User interface development during the HF process[a]

HF activity	Example inputs to design	Considerations for physical device	Considerations for labeling
User profiles	Previous experience with similar products	Align functionality with commonly used products Avoid drastic deviations from current procedure	Use IFU to draw attention to key design differences Consider on-device labeling or quick reference guide for major deviations from common procedure
	Lack of product experience	Simplify procedures	Provide detailed descriptions that limit assumptions about user knowledge Avoid jargon
	Physical user characteristics	Accommodate the lower end of physical capabilities Avoid the need to manipulate small parts, requiring high dexterity	Include details of such techniques (e.g., depress plunger slowly) or time to complete tasks so that users know exactly what physical effort to expect
	Cognitive user characteristics	Limit the number of steps required to use the product Develop procedures that promote memorability	Use simplified instructions and terminology Create easy-to-navigate menus Utilize graphics to complement text
Profile of use environment	Physical conditions (e.g., light, noise, temperature, space constraints)	Utilize both audio and visual cues for important device functions Design digital screen to limit glare and improve contrast	Incorporate large fonts and graphics for poorly lit environments Consider booklet designs or digital IFUs for complex devices used in space constrained environments Specialized materials may be needed for use in a sterile environment
	Storage location	Minimize product footprint to maximize storage possibilities Utilize strategies to limit drug exposure to light (combination products)	For IFUs that may be retained by the user (e.g., reusable products), consider both durability and printing on the outside of the IFU for easy identification
	Computer or digital access	Design to accommodate varying platforms of use (e.g., web access, apps) Limit need for Wi-fi connection for product to function accurately	Identify the need for digital IFU based on likelihood it can be accessed at the point of use Design of IFU may need to be modified to accommodate viewing experience on digital platform

(continued)

Table 6.5 (continued)

HF activity	Example inputs to design	Considerations for physical device	Considerations for labeling
Task analysis and use error analysis	Product use scenarios	Indicating clear completion of use scenarios and that users have achieved their goal	Use this to determine IFU scope and sections
	Task steps	Identify potential controls to promote the correct task order	Task steps drive order of step-by-step instructions
	Failure modes (use errors)	Opportunities to prevent failure modes or address causes of failure	Helps identify warnings/cautions related to safety
	Critical task identification	Opportunities to eliminate harmful outcomes or reduce probability of high-severity harms	Prioritize information to emphasize in the labeling Identify information that may require multiple labeling mitigations (e.g., IFU, on-device labeling, information printed on packaging)

[a][1] AAMI TIR49: 2013 - Technical Information Report - *Design of training and instructional materials for medical devices used in non-clinical environments.* [2] Gernazian, Valerie; Pandey, Abhinita. "Integrating IFU Development with 4 Key Elements of the HFE Process." Agilis Consulting Group, Agilis Consulting Group, 7 Jul. 2023

Conclusion

Throughout this chapter, we have touched on aspects of regulation impacting patient safety and highlighted some key considerations for developers. Rigorous technical work and a strong commitment to the scientific process are imperative to achieve regulatory success, but meeting only the minimum bar for market entry is unlikely to result in an optimized product, which could lead to poor user satisfaction and even deliberate misuse. Making intelligent product decisions demands a true understanding and empathy, not only for patients receiving the treatments we develop, but for anyone who may interact with drug products or operate devices to deliver treatment. In some cases, ensuring that the end users receive the best product could result in longer development timelines and added costs, but the trade-off of producing a product that provides a safer and more enriching experience for users may be worth it. Though these considerations should always be at the forefront of any development activity, trade-offs are often necessary to get well-made products to market quickly. By putting the same emphasis on early stage and formative activities that are not required, developers can produce superior products with a greater chance of success. Regulators keep industry in check, but it ultimately falls on the manufacturer to make the right choices.

Chapter 7
Diabetes Care Case Study

Sherri Biondi, Colleen Foley-Broich, Trang Ly, Wolfgang Fraunhofer, Ramakrishna Venugopalan, and Robert R. Nesbitt

Abstract "This chapter is a case study outlining at a high-level how changes in technologies in diagnostics, therapeutics, and delivery systems have addressed patient needs in diabetes care over the past 100 years."

Keywords Patient Self-Blood Glucose monitoring (sbgm) · Pen-Injectors · Continuous glucose monitoring (CGM) · Automated Insulin Delivery (AID) · Noninsulin therapies

Throughout this book, we have seen examples of conditions in which the patient has the challenge of integrating a lot of information and making critical decisions about their care. They need to understand diagnostic or biomarker data, determine how much medication to take and when is the right time to take it.

S. Biondi (✉)
Device Development, Dosage Form Design and Development, Biopharmaceuticals R&D, AstraZeneca, South San Francisco, CA, USA
e-mail: sherri.biondi@astrazeneca.com

C. Foley-Broich (✉)
Marketing Research/Commercial Insight, Abbvie and Horizon (Amgen) Therapeutics, Chicago, IL, USA

T. Ly (✉)
Clinical and Medical Affairs, Insulet Corporation, Acton, MA, USA
e-mail: tly@insulet.com

W. Fraunhofer
Drug Product Development, AbbVie, Inc., North Chicago, IL, USA
e-mail: Wolfgang.fraunhofer@abbvie.com

R. Venugopalan
Continuity Biosciences LLC, Malvern, PA, USA

R. R. Nesbitt
Combination Product Development and Drug Delivery, AbbVie, Inc., North Chicago, IL, USA

R. R. Nesbitt et al. (eds.), *Patient-focused Development of Biopharmaceutical Products*, AAPS Advances in the Pharmaceutical Sciences Series 27,
https://doi.org/10.1007/978-3-031-98253-8_7

Regardless of disease, patients are challenged to take what they learn about their disease and their therapy and find a way to make disease management fit in their life as effectively and seamlessly as possible. One of the conditions to which we have referred is diabetes (looking at both type 1 and type 2 diabetes). In this chapter, we would like to look—at a high-level—at how patient needs have driven (or been impacted by) developments in technology in managing diabetes.

Diabetes—what we formally call *diabetes mellitus*—has been a challenge for mankind for millennia. The term "diabetes" was first used by Apollonius of Memphis around 250–300 B.C. [1]. A little over 100 years ago, insulin was first isolated and shown to be an effective treatment for *diabetes mellitus*. In those 100 years, those with diabetes have had to find an appropriate set of tools to understand the state of their health (i.e., by measurement of their glucose levels) and from that data (combined with other input) influence their diet and determine how much insulin (or other noninsulin therapies) to utilize to manage their disease. In this case study chapter of patient needs in treating diabetes, we will focus on the two major types within the broader category of *diabetes mellitus*—type 1 (in which the pancreas stops producing endogenous insulin) and type 2 (in which the pancreas either does not produce enough insulin, and/or the body is unable to utilize insulin as previously [2]). (For simplicity, we will not address gestational diabetes or other more recent research into different variations). To help understand the tools with which patients manage their disease, we will separate their needs into three components:

1. *Diagnostic*: How the patient can assess their body's glucose level (e.g., blood glucose monitor).
2. *Therapeutic*: The drug or biologic to treat the disease in the patient's body (i.e., insulin, Glucagon-Like-Peptides [GLPs]), and,
3. *Delivery Device*: The system by which the drug is administered (e.g., pen-injector, pump).

A high-level view of major development in each of these three areas is shown in Fig. 7.1.

Regardless of the state of technology in each of these areas, the consistent challenge for patients (and their clinicians) has been finding the right balance of timely, actionable data (the diagnostic), the appropriate therapeutic, clarity on how much to administer (and when) and finally, the appropriate way to administer the drug. Over many years, development in each area provided patients with improved diagnostic, therapeutic and delivery options but more importantly, innovations enabled patients to integrate these three areas to manage their disease more simply and more effectively:

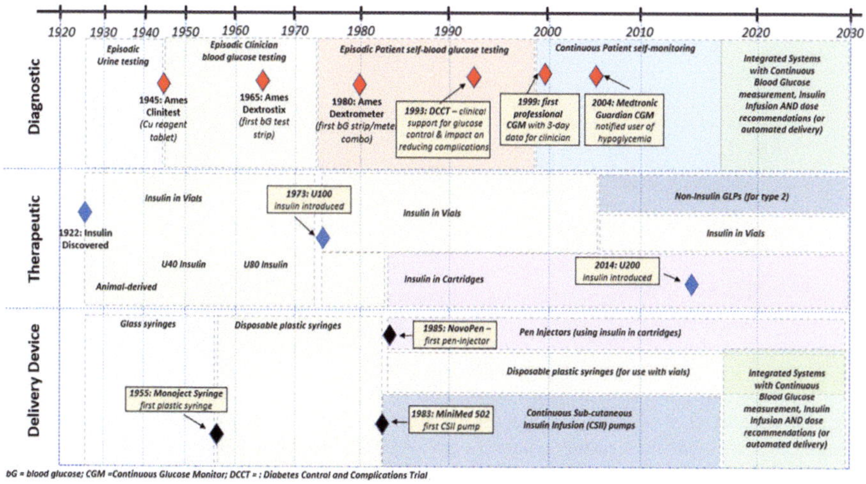

Fig. 7.1 Diabetes treatment through the years

Diagnostics

Before the advent of modern continuous blood glucose meters, the management of diabetes relied heavily on indirect methods of monitoring. In the early 20th century, urinalysis was most commonly used to provide a view of diabetic status. While this tool had many limitations, it was incredibly useful in its day and helped guide dietary adjustments and insulin dosing (once available). The challenge was that glucose appeared in the urine only when blood sugar levels surpassed a certain threshold and so mild hyperglycemia could go undetected. Additionally, since the measurement was episodic, it reflected the accumulation of sugar over time and did not provide an accurate real time measurement. The first significant shift in diagnostics was moving from urinalysis to blood glucose (bG) measurement and moving from infrequent clinician bG measurement to more frequent, at-home patient self-blood glucose monitoring (sbgm), with obvious advantage to the patient. Advances in bG measurement enabled a shift from colorimetric (matching the color on a strip) to electrochemical measurement, which enabled more accurate measurement with progressively smaller blood samples required (and less pain).

Although moving from urine analysis to blood glucose measurement was easier for patients to use, either one still only gave a "snapshot" of their health. These episodic snapshots did not give the patient any indication if their blood glucose was increasing or decreasing—nor did it indicate "how fast."

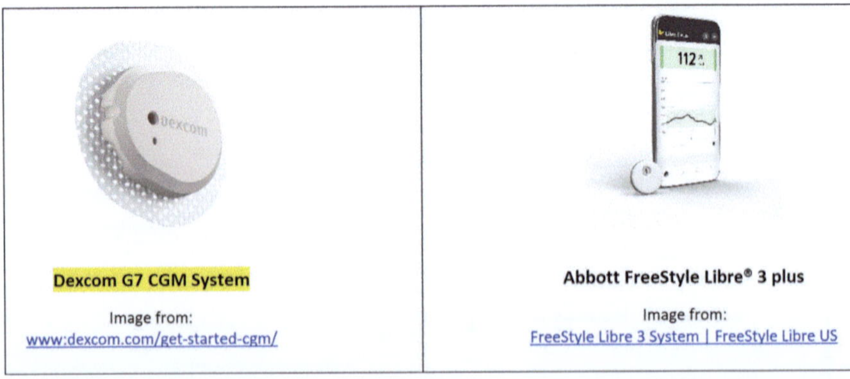

Fig. 7.2 Examples of continuous glucose-monitoring systems. (**Dexcom G7 CGM System**. Image from: www:dexcom.com/get-started-cgm/. **Abbott FreeStyle Libre® 3 plus**. Image from: FreeStyle Libre 3 System I FreeStyle Libre US)

Continuous Glucose-Monitoring Systems

The development of continuous glucose monitoring (CGM) systems has proven to be a transformative technology for people with diabetes. This technology consists of a device that continuously tracks glucose in the interstitial fluid using sensor inserted under the skin. This data is typically sent wirelessly to a receiver such as a smart phone app, a dedicated monitor or an insulin pump and allows the wearer to see (and be notified about) blood glucose trends in direction and rate to be able to take action to prevent (or minimize) hypoglycemic (low blood sugar) or hyperglycemic (high blood sugar) events. This is very helpful for type 2 patients, but especially crucial in managing type 1 diabetes. And although this information is vital for adults, it is an especially valuable tool for parents of pediatric type 1 patients who are generally unable to do the required monitoring or be aware of the consequences. These CGM systems have been hugely beneficial in enabling parents of these children to sleep at night, with the confidence that these systems can remotely notify them of changes in bG levels, especially rapid rates of change. Examples of these systems are shown in Fig. 7.2:

Therapeutics

Before the 1920's, diabetes could be a life threatening disease. In 1921 scientists succeeded in extracting and purifying animal-based insulin and demonstrated it could be effective at lowering blood sugar. Advances in insulin over the first 50 years from its discovery and availability, enabled a move from animal-based insulin to human-based insulin, reducing incidents of immune responses.

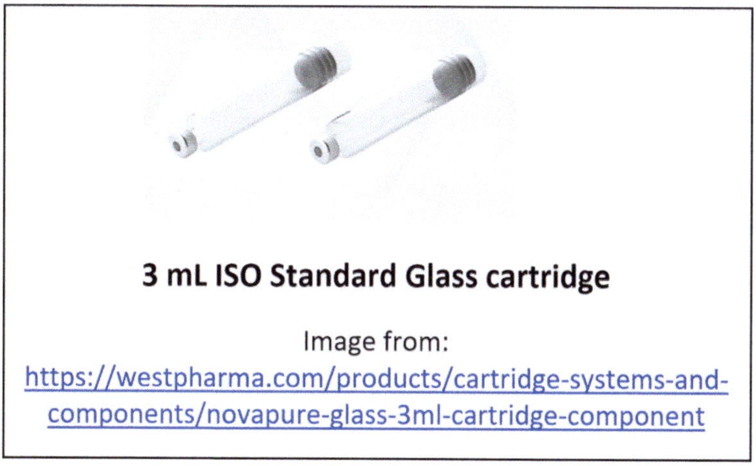

Fig. 7.3 Examples of 3.0 mL ISO Standard Glass cartridges. (Image from: https://westpharma.com/products/cartridge-systems-and-components/novapure-glass-3ml-cartridge-component)

Insulin was first marketed as 5 units per milliliter (U5) with higher concentration formulations following soon after (U10, U20, U40, U80). In the 1950's this was standardized to U100 concentration with lower concentration products becoming mostly obsolete [add reference to Sims et al, "100 years of insulin: celebrating the past, present and future of diabetes therapy", 2021] The move from a variety of previous insulin concentrations (U40 or U80) to a standard concentration (U100) reduced the risk of confusion and any accompanying hypoglycemic risk. In recent years, manufacturers developed and commercialized several new diabetes drug products to meet the diversity of patient administration needs. First, higher concentration insulins (U200 and U500) were commercialized for type 2 patients with significant insulin resistance or high insulin requirements. This reduced the total dose volume for these patients compared to when they were dosing from U100 formulations. Additionally, new classes of anti-diabetes therapies such as GLPs (glucagon-like peptides) have been developed that provide benefits to those with type 2 diabetes in reducing bG levels with less risk of hypoglycemia (and less frequent injections).

Moving into the late twentieth century, while diabetes therapeutics matured, the drug product primary container also evolved. Initially insulin was commercialized in ampules and vials, and required patients to draw and set the dose correctly with a syringe. In the late 20th century, manufacturers began commercializing insulin in 1.5ml and 3.0ml glass cartridges to support use with pen injectors (discussed in the next section). These options provided a significant improvement in usability and safety for patients as well as a diversity of choice to meet individual needs and preferences for self-administration.

Example of 3.0-mL ISO Standard Glass cartridges are shown in Fig. 7.3.

Delivery Devices

In the last 75 years, there have been sgnificant advances in the devices that deliver therapeutics to treat diabetes and these have meaningfully improved daily life for patients.

Syringes for patient self-injection of insulin were first available in the 1940s and manufactured by Becton Dickinson and Company (BD, Franklin Lakes, NJ). (add ref Heinemann et al "Needle technology for insulin administration ..." 2021) They were intended for the low concentration insulins available at the time, so large doses were often required. As these devices were constructed of glass and stainless steel, they had to be boiled after each use to keep them sterile. In addition, the reusable syringe needles required periodic sharpening by patients.

The first plastic insulin syringe was available in 1954, which provided patients the opportunity to take their daily insulin injections via one-time use, disposable syringes. The use of these devices has provided additional convenience for patients. Because disposable syringes doid not require sterilization, patients had significantly fewer complex preparatory steps prior to injecting which reduced errors and improved safety considerably. An example of a disposable insulin syringe is shown in Fig. 7.4.

Pen-Injectors: A significant improvement in usability, discretion and portability was achieved with the invention of pen-injectors, which allow patients to dial up the required dose amount and discretely take their injection without the need to withdraw insulin with a syringe from a vial. This enabled multiple daily doses of insulin to be administered easily and discretely.

In 1985, Denmark's Novo Terapeutisk Laboratorium (which merged with Nordisk Insulinlaboratorium in 1989) created a new class of product with the launch

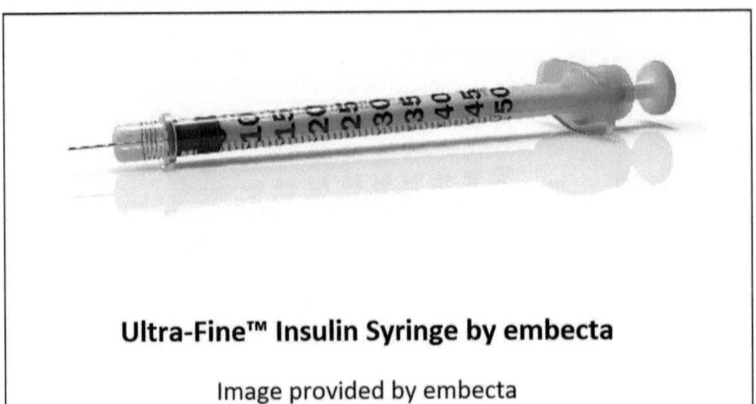

Fig. 7.4 Example of disposable insulin syringe. (Ultra-Fine™ Insulin Syringe by embecta Image provided by embecta)

of the NovoPen®. This device was a major advancement in portability and discretion for patients requiring insulin. The device contained a 1.5-ml glass cartridge of insulin and utilized a removable needle—all contained in a device that resembled a writing pen. Although this first pen-injector did not have a dial to display the dose to be delivered (as current ISO standards require) and administered 2 units (20 microliters) of insulin with each depression of the plunger, it was a significant breakthrough and paved the way for the class of insulin pen-injectors, which are widely used today.

Like the NovoPen®, initial pen-injectors were reusable devices in which the insulin cartridges could be removed and replaced when emptied. Generally, these reusable devices are mechanical devices, which display the dose that the user dials up on the knob at the end via a dial (often on the side of the device).

Some pen-injectors keep a memory of the past doses delivered, such as the Novo Nordisk® Echo® (shown in Fig. 7.5). These devices help the user keep track of how much insulin was delivered, and how long ago. This can be very helpful in a multidaily insulin delivery therapy where every day can blend into the next, and mid-morning (for example), the patient can forget whether they really took their insulin in the morning (or was that yesterday?).

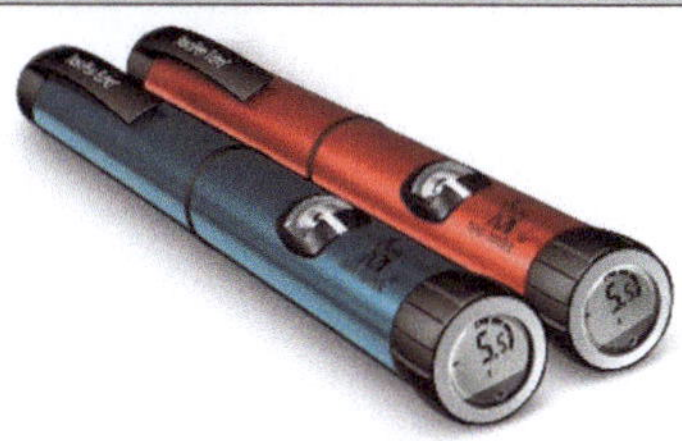

Novo Nordisk NovoPen® Echo®

Image from: www.novonordisk.com/our-products/pens-and-needles/novopen-echo.html

(Image approved for use by Novo Nordisk)

Fig. 7.5 Examples of refillable insulin pen that can record doses. (**Novo Nordisk NovoPen® Echo®** Image from: www.novonordisk.com/our-products/pens-and-needles/novopen-echo.html (Image approved for use by Novo Nordisk))

Fig. 7.6 Example of prefilled, disposable insulin pen. (Lilly **Humalog® KwikPen®** Image from: https://uspl.lilly.com/humalog/humalog.html#ug1 © Copyright Eli Lilly and Company. All Rights Reserved. Used with Permission)

In 1989, Novo Nordisk launched the first disposable, prefilled pen-injector, called the NovoLet. Unlike reusable insulin delivery devices ("pens") in which the insulin cartridges could be removed and replaced when emptied, prefilled insulin pens had the insulin cartridge permanently integrated, and—when the cartridge was empty—the pen could be disposed. Patients preferred the simplicity of these prefilled devices, which eliminated the cartridge replacement steps. The first prefilled pen on the market, the NovoLet, required the user to look at the dose numbers on the center of the pen—and keep track of the number of revolutions to know the dose dialed. Although users liked the simplicity of a prefilled pen, the dose setting methodology was, for some, confusing.

The next generation of prefilled insulin pens eliminated any potential dose-setting confusion. Each of the major insulin manufacturers (Novo Nordisk, Eli Lilly and Company, and Sanofi) launched simple, easy-to-use prefilled insulin pens with similar functionality—turn the dial to display the dose to be delivered, and the dial "dials down to zero" as the dose is delivered. An example of such a prefilled insulin pen is shown in Fig. 7.6.

Insulin Pumps

Insulin pumps are portable battery-operated devices that deliver insulin from a reservoir via a needle inserted into the subcutaneous space. Insulin pumps first became available in the 1970's but became more widely used in the early 2000's when clinical data demonstrated the value of tight glycemic control on minimizing diabetes-related health complications (add ref, Berget et al, A clinical overview of insulin pump therapy for the management of diabetes, 2019). Today, more than 60% of Type 1 diabetes patients use pumps to manage their diabetes due to the superior glycemic control and the reduced burden of care (e.g. minimizing multiple daily injections and providing discrete therapy management) (add ref: Gandhi et al, "Insulin pump utilization in 2017-2021 ..."). Over these many years insulin pumps have evolved and improved significantly in many patient-centric ways. Early pumps were large and heavy with many preparation steps and a complex user interface that people often found difficult to program. Manufacturers developed and commercialized many generations of improvements to these early pumps – reducing the size and weight and simplifying the user interface while simultaneously enabling complex personalized basal and bolus dosing regimens that suited individual patient needs.

Today, patients have a choice between many insulin pumps (also called CSII or Continuous Subcutaneous Insulin Infusion systems) depending on their needs. Most are belt-worn devices where insulin is delivered through tubing connected to an infusion set. Patients can adjust delivery parameters directly on the device, and for many pumps, make adjustments via an external monitor or mobile phone app. An additional class of insulin delivery systems was created with the development of disposable insulin delivery pods. In 2005, Insulet (Acton, MA) received FDA clearance for the Omnipod® which provided a significant advancement in insulin delivery for type 1 diabetes patients. The Omnipod differed from traditional CSII systems such as those described above, in that it (a) delivered insulin directly through a tubeless pump into the subcutaneous tissue below the adhesive-backed "Pod" (which contains the insulin), and (b) was controlled by a remote controller rather than a display and user interface on the surface of the pump as seen on other CSII systems.

This afforded a level of discretion and portability that was not possible with the smallest traditional CSII pump. By eliminating the tubing and separate infusion port, only the "Pod" was visible, and this could be placed on the abdomen or any location that an injection could be given. An image of the Insulet Omnipod DASH is shown in Fig. 7.7.

Automated Insulin Delivery (AID) Systems

What we have described to date are independent and separate blood glucose-monitoring systems and insulin delivery systems which required patients to interpret their glucose data and then make decisions about their insulin dose and then deliver that dose accordingly. This created a great deal of cognitive and psychological burden for the patient as they manually connected information between the two. In the discovery research described in Chap. 1, we heard patients with type 1 diabetes describe all the things they need to think about in their day-to-day life in dealing with diabetes and their feeling that this creates a "white noise" [5] in their mind—the ever-present thought about managing their disease—all invisible to others.

A significant advancement in diabetes care has been created in recent years, which can automatically integrate the independent blood glucose value and the amount of insulin that is required—via what is called Automated Insulin Delivery (AID) systems. This is also referred to as *Sensor-Augmented Therapy* [4]. These systems have complex algorithms which analyze glucose data, make recommendations on how best to adjust insulin delivery, and then automatically modify insulin

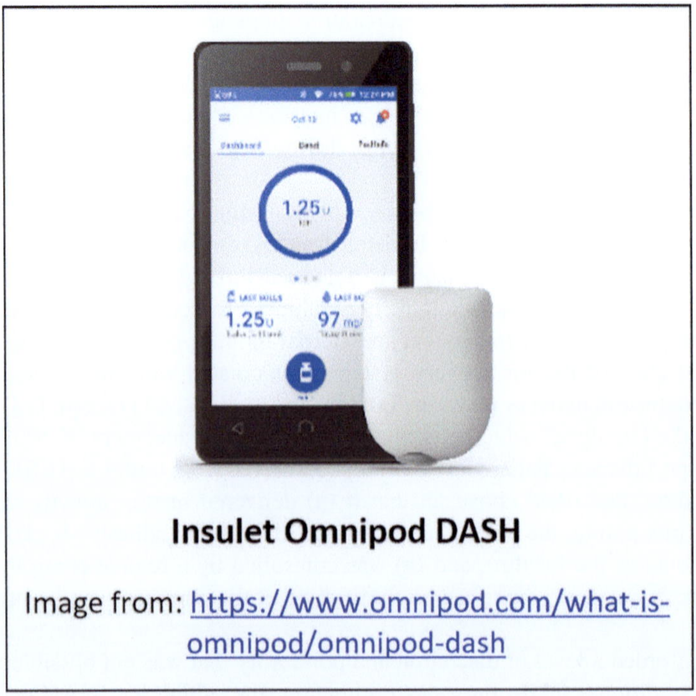

Fig. 7.7 Example of a CSII insulin pump. (**Insulet Omnipod DASH** Image from: https://www.omnipod.com/what-is-omnipod/omnipod-dash (Note: Pod is shown without the necessary adhesive)

pump settings to achieve the desired glycemic control. (add ref, Ware et al, "Close-loop insulin delivery: update on the state of the field and emerging technologies", 2022).

Why it Is Important for Patients: A Case Study

As an example of why these AID systems can be of such great value, perhaps a case study will be instructive:

Maria is a 60-year-old woman who has had type 1 diabetes for 50 years. She has had hypoglycemia unawareness (i.e., when a person with diabetes is no longer able to sense when their blood sugar is getting low) for several years, putting her at risk for severe hypoglycemia. She has had moderate deafness and often does not wake up to her CGM alerts overnight. Maria's doctor would like her glucose levels to be 130–180 mg/dL overnight to reduce her risk of hypoglycemia. During the daytime, Maria would like more aggressive glycemic management than recommended during the nighttime.

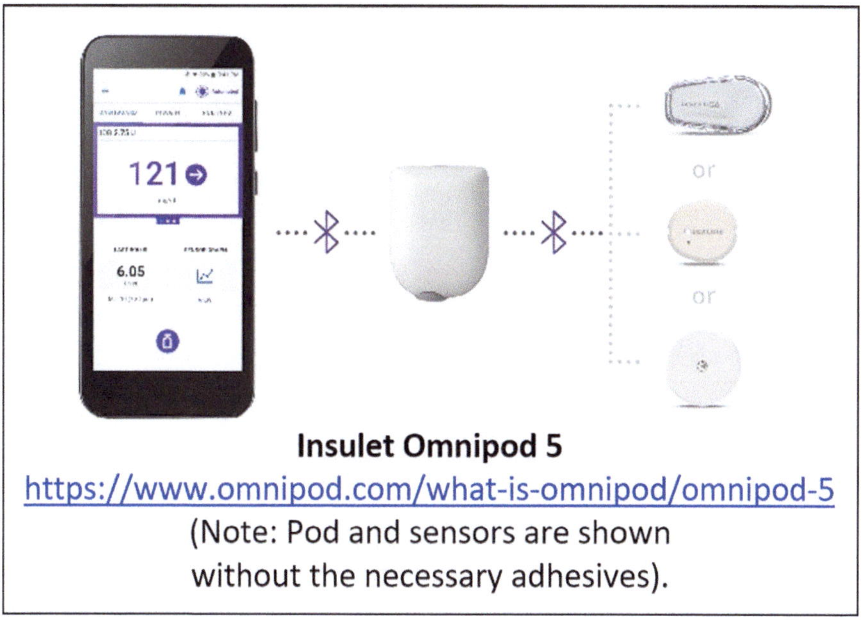

Fig. 7.8 Example of an Automated Insulin Delivery (AID) system. (**Insulet Omnipod 5** https://www.omnipod.com/what-is-omnipod/omnipod-5 (Note: Pod and sensors are shown without the necessary adhesives))

The benefit that Automated Insulin Delivery systems can provide patients like Maria (and even more, pediatric Type 1 diabetes patients and their families) is the ability to automatically integrate her real-time, continuous glucose sensor input and deliver the right amount of insulin appropriate at that time.

For Maria, her Continuous Glucose Monitor (CGM) sensor sends glucose values and trends to her OmniPod to help the system make automated insulin delivery decisions. FIgure 7.8 below shows the Omnipod® 5 Automated Insulin Delivery System and how it pairs with the Dexcom G6 or G7 CGM System or the FreeStyle Libre 2 Plus Sensor.

For Maria, the Omnipod 5 System allows for dynamic insulin adjustment every 5 minutes based on real-time continuous glucose sensor input. It takes accurate sensor values and delivers a personalized insulin dose to meet her needs. Moreover, the Omnipod 5 System also allows her to personalize her insulin delivery by choosing the target glucose. This is the value the algorithm aims for when automating insulin delivery.

In this case, because of Maria's hypoglycemia unawareness and concern about not waking to alarms overnight, her physician sets her overnight glucose target to 150 mg/dL to provide more protection against nocturnal hypoglycemia. During the day, Maria's physician sets the target lower, to 120 mg/dL. The clinician knows she can set different target glucose values for specific times of day to optimize

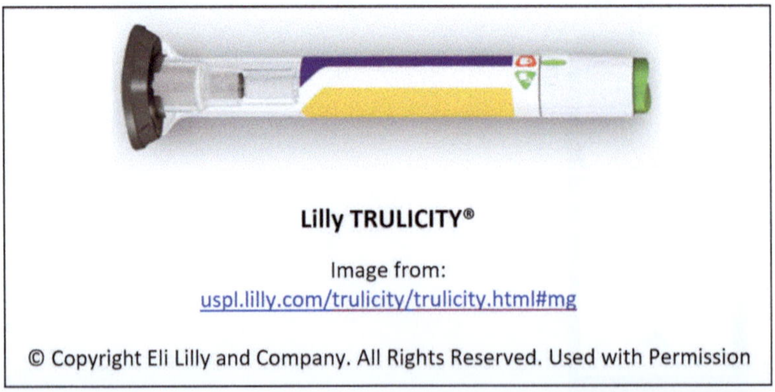

Fig. 7.9 Example of a GLP injection device. (**Lilly TRULICITY®** Image from: uspl.lilly.com/trulicity/trulicity.html#mg © Copyright Eli Lilly and Company. All Rights Reserved. Used with Permission)

glycemia. Therefore, she explains to Maria that these targets can be adjusted as needed across time to meet her personal diabetes management goals.

These advances in automated insulin delivery have been driven in large part by the advocacy of the diabetes community with companies and regulators. In addition to algorithm advancement, future innovation will be focused on making devices last longer, further reducing the size of devices to minimize skin real estate, and making the user interface easier to use. This will reduce the barriers for adoption so that more people can access this technology to manage their complex insulin therapy.

Noninsulin Therapies

Although so far, this chapter has focused on insulin as a diabetes treatment, there are a variety of other therapeutics to treat diabetes. More traditional treatments have included oral antidiabetic medications (e.g., metformin, sulfonurea) used to initially manage bG levels. However, over the past 20 years, another class of noninsulin therapeutics has been developed that can enable patients to get the benefit of the blood glucose-lowering capabilities of insulin without the fear of hypoglycemia.

This class of therapeutic includes "GLPs" or glucagon-like peptides, which mimic the action of incretin glucagon-like peptide-1 to manage diabetes type 2. These GLPs can be provided in an oral dose form (e.g., semaglutide, RYBELSUS™, by Novo Nordisk) as well as subcutaneous injection (e.g., semaglutide, OZEMPIC™, by Novo Nordisk, or dulaglutide, TRULICITY™, by Eli Lilly and Company). An example of Trulicity, is shown in Fig. 7.9.

This class of therapeutics is analogs of the human hormone with additional attached molecular structure portions to optimize the uptake and metabolism of the active peptide. These structural arrangements allow for a much longer half-life, extending the time required between successive doses. These subcutaneous

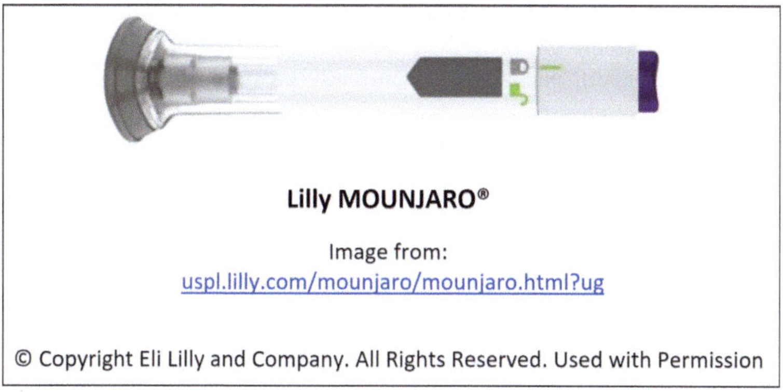

Fig. 7.10 Example of a GIP-GLP injection device. (**Lilly MOUNJARO®** Image from: uspl.lilly.com/mounjaro/mounjaro.html?ug © Copyright Eli Lilly and Company. All Rights Reserved. Used with Permission)

injections are generally weekly and lessen the burden of managing diabetes dosing schedules as compared to insulin-based therapies requiring daily dosing, as the weekly dosing format reduces the number of injections required relative to insulin delivery for type 2. The single full dose delivery of an autoinjector removes concerns of errors in dialing up an incorrect dose of a variable dose delivery device. However, the use of a single-dose autoinjector does increase sharps disposal waste.

More recently, another new class of diabetes therapeutics has been introduced, which add an additional hormone to the glucagon-like peptide 1 (GLP-1) in treating diabetes—these are called "GIP-GLPs," as the therapeutic is a combination of glucose-dependent insulinotropic polypeptide (GIP) receptor and glucagon-like peptide-1 (GLP-1) receptor agonist. An example of this therapy is tirzepatide (MOUNJARO™, Eli Lilly and Company), recently approved for diabetes type 2 patients (shown in Fig. 7.10). The chemical structure has been modified as an analog of human GIP, with a fatty-diacid portion attached to the polypeptide, increasing the half-life of the therapeutic within the blood stream due to high affinity to albumin, and thus greatly extending the time required between doses. Delivery of a therapeutic with long duration between dosing might be best administered by autoinjector to lower any dosing errors due to forgetting complete dosing schemes between injections.

Conclusions and Forward-Looking Thoughts

Diabetes is an interesting disease to consider, and one we wanted to highlight in this book, because of the long history of diagnostics and therapeutics as well as the unique and complex needs of the patient population. At the start of this chapter, we noted that effective disease management requires a complex interplay between three things—the continual diagnostic assessment of glucose levels, the selection of the right therapeutic to manage the disease, and the safe and effective use of the delivery device to administer the required dose of the drug. Over many decades, a deep understanding of patient needs has driven really meaningful improvements in all three areas and provided benefits that patients and health-care providers value. These include the benefits of less frequent finger sticks to measure blood glucose, more intuitive data collection, analysis, and communication to facilitate better and more timely decision making, more efficacious therapeutics—many of which enable less frequent dosing, and delivery devices with more intuitive device user interface designs, which facilitate safe and effective administration. Most notably, improvements in this space recently have included the integration of multiple devices, to enable diagnostic measurements, data analysis, and accurate dosing all in a single device. This is, perhaps, the most significant advancement in patient centricity yet.

Diabetes, more than other disease areas, has seen many generations of technological improvement that benefit the unique needs of the patient population. An understanding of this long history and the needs/perspectives of this patient population can provide insights to developers of next-generation products in this space as well as developers designing products in adjacent disease areas, as they consider what patient-centricity means to them.

References

1. Sapra A, Bhandari P, Diabetes Mellitus, StatPearls Publishing, Treasure Island, FL, 2022.
2. Some references which describe Diabetes type 1 and 2.
3. Alsaleh, F.M., Smith, F.J. et al, Insulin Pumps: from inception to the present and toward the future, 5 March, 2010, Journal of Clinical Pharmacy and Therapeutics, Vol. 35, Issue 2, April 2010.
4. Sensor-Augmented Therapy (SAP) is defined in Kesavadev, J et al, Evolution of Insulin Delivery Devices: From Syringes, Pens and Pumps to DIY Artificial Pancreas, Diabetes Therapy, 14 May 2020,
5. Reference to ethnographic research done at Lilly in early 2000's

Chapter 8
Now What?

Sherri Biondi, Colleen Foley-Broich, Trang Ly, Ramakrishna Venugopalan, and Robert R. Nesbitt

Abstract This chapter shares some key observations, which might help guide the reader in ensuring a thorough understanding of the patients' needs and means to ensure those are built into the development of a biopharmaceutical product.

Keywords User diversity · Intuitive design · Understanding the competitive environment · Patient stories

We have covered a lot of ground in this book. We have sought to help the reader understand patient needs from two perspectives. First, considering the patients themselves – and understanding their needs, their ecosystem and their lived experiences. The second perspective is that of the designer or product developer and understanding how somewhat ambiguous patient needs translate into design requirements and ultimately the product itself.

We have also sought to ensure the reader as a developer of these products understands the wide range of seemingly conflicting trade-offs a firm must navigate, and

S. Biondi (✉)
Device Development, Dosage Form Design and Development, Biopharmaceuticals R&D, AstraZeneca, South San Francisco, CA, USA
e-mail: sherri.biondi@astrazeneca.com

C. Foley-Broich (✉)
Marketing Research/Commercial Insight, Abbvie and Horizon (Amgen) Therapeutics, Chicago, IL, USA

T. Ly
Clinical and Medical Affairs, Insulet Corporation, Acton, MA, USA
e-mail: tly@insulet.com

R. Venugopalan
Continuity Biosciences LLC, Malvern, PA, USA

R. R. Nesbitt
Combination Product Development and Drug Delivery, AbbVie, Inc., North Chicago, IL, USA

R. R. Nesbitt et al. (eds.), *Patient-focused Development of Biopharmaceutical Products*, AAPS Advances in the Pharmaceutical Sciences Series 27,
https://doi.org/10.1007/978-3-031-98253-8_8

the kinds of considerations that regulators will use to evaluate their product, as well as how the Commercial or Marketing team needs to find the right balance of promotion and support of these products in a highly competitive but regulated space.

Some of the key observations we would like to highlight

1. *Users are diverse*: Users have a range of lived experiences and therefore bring a surprisingly broad array of expectations and experiences when they sit down to use a product. Most importantly, users are very likely to approach product use very differently from you, the developer, and so a deep understanding of the user—their journey, the context they bring to the table, as well as their physical and cognitive capabilities—will be key in understanding good product design. In particular, we want to emphasize the need to consider the full context of the product's use prior to, and throughout, the development of the product. A deep understanding of the patient's ecosystem that informs their needs—as well as those of caregivers and health-care professionals—should be the touchpoint of developing and testing these devices.
2. *There are many lenses by which to understand their needs*: We describe a variety of different lenses through which a developer can seek to understand users and their environments—whether by direct observation, shadowing, diaries, direct questions, or other methods. This data collection yields information through which developers can start to build an understanding of user needs. It is critical to carefully consider which method to utilize to gather the right information at the right time. Though this work typically begins near the start of development—during device ideation stages and before prototypes have been designed and tested—it may continue throughout the product development lifecycle and is customized based on specific product requirements.
3. *The development process is complex*: We describe the drug and combination product development process to help the reader appreciate the complexity of CMC development along with clinical studies. It is our hope that the reader has gained an appreciation for how human factors engineering fits into this process. Typically, once a product is advancing to clinical stages of development, the device design becomes mature enough so that formative user testing can be conducted to evaluate product usability and use-safety in the hands of representative users. At this stage, use-related risk analyses (URRAs) become fleshed out in detail, facilitating the identification and analysis of risks that might impact misdosing or other safety-related risks in clinical studies, which will eventually translate to the market.
4. *We need to think about the regulatory expectations throughout development*: We have provided readers with both a regulatory and commercial perspective in this space. From a regulatory perspective, global health authorities can be quite prescriptive in submission requirements and the data required to support safe and effective use. An up-front knowledge of these requirements ensures the necessary data is collected throughout development, since it will be critical for approval.

5. ... *and understanding the competitive environment is vital:* We have provided a commercial perspective so that the reader can consider what will be required to prepare for and support launch and ongoing supply of a product, as well as how to deal with the competitive pressures that will surely emerge during the product's life.
6. *How can we make it intuitive*: Not only should we be designing the best product for delivery of therapeutics as the patient manages their health-care needs, but we must also provide a product that is intuitive and overcomes any patient preconceived notions /expectations of how a product is to be used(e.g. wouldn't it be nice if we could use the simple 0.1 s dose injection into the neck we saw on Star Trek!). A product designed only to meet regulatory expectations for safety will not necessarily be one that is easy or satisfying for the intended user population.
7. *Patient stories speak volumes*: As you've seen, throughout the book, we've provided a rich collection of patient stories—their journeys before and after diagnosis and their challenges onboarding onto new therapies. These journeys are often bumpy ones where the complexity of the health-care system, their diagnoses, their diseases, and their personal lives interfere in surprising ways with the safe and effective use of drug delivery devices. Just as we have used these stories to facilitate understanding of development considerations in our book, remember that sharing (appropriately anonymized) stories of patients' needs also helps your development team stay connected to the "North Star"—which is to serve the needs and expectations of our patients.

Our intention in creating a chapter aimed at summarizing 100 years of products to support patients with diabetes mellitus was not to provide an exhaustive or complete survey, but rather to illustrate how over time, various industries have supported the needs of these patients—first by supplying individual components of a treatment regime (therapeutics (insulins, and noninsulin GLPs), diagnostics, and drug delivery systems)—and, over time—evolving to provide more integrated solutions that provide a better experience and ultimately more effective therapy. We felt this case study could be illustrative of this balance of understanding not just the technical effectiveness of any individual component (e.g., improving the clinical accuracy of a discrete blood glucose measurement) to evolving to development of integrated systems, which meet the real-life needs of the patients (e.g., integrated continuous glucose-monitoring systems (CGMSs), which use an algorithm to recommend and (upon confirmation) can deliver appropriate insulins.

Spend the time investing in understanding your patients' needs and keep this visible and regularly touch back on that understanding throughout the development, and you can be more confident your product will meet their needs and be an important (but appropriately invisible) part of their life.

Glossary of Terms

Term	What It Means to Us in This Book
Caregiver	This is any person who supports the management of the patient's disease, which could include a parent, spouse, child but can also include an interested relative, neighbor, or any other (non-HCP) who is engaged in supporting the administration of the therapy.
Combination product	A product comprised of two or more regulated components - drug, device or biologic - that are combined into a single entity, co-packaged as a unit, or separately packaged but intended for use only with each other.
Connected care or connectivity	We use these terms to mean the use of apps, smartphones, or iPads to augment the treatment of the disease.
Dosing	The schedule or regimen by which the therapeutic is to be administered to achieve the intended medical effect. This is usually expressed in frequency over time (e.g., twice daily, once daily, once every other month).
Ecosystem	The environment in which the patient operates – not just their home and work, but how they interact with health-care professionals and how they access their medications and medical devices (including pharmacies).
Formulation	The combination of ingredients, which will achieve the intended effect in the recipient, including, but not limited to, the active drug product, as well as other substances to enhance the active ingredients' absorption or stability.
Health-care professional (HCP)	We use this to mean any range of clinical support, from a nurse to a doctor. It could also include therapists used in accommodating symptoms of the disease (occupational therapists, physical therapists, etc.).
Patient journey	What the individuals being treated for a disease generally experience through the progression of their disease. It includes early symptoms, diagnoses, and treatments; how the disease progresses (slowly or suddenly); how the disease manifests itself; and what range of treatments are used throughout the progression. Although we use the term "patient," these journeys can additionally address how the caregivers and health-care professionals who support these patients react to and are involved in the support of their loved one. These are often complex images, which address practical information (standard-of-care drugs and treatments) as well as emotional and psychological implications of the disease on their life journey.

R. R. Nesbitt et al. (eds.), *Patient-focused Development of Biopharmaceutical Products*, AAPS Advances in the Pharmaceutical Sciences Series 27,
https://doi.org/10.1007/978-3-031-98253-8

Term	What It Means to Us in This Book
Platform	Generally, an architectural structure, which provides the basis for a variety of iterations. In our book, we use this term to refer to a delivery device, which can be configured or deployed in a variety of similar manners to support different drug deliveries to support different disease states.
Product presentation	We use this term to represent the format in which the drug is provided to the patient to treat their disease. These can include solid-dose form (tablet or capsule) for oral delivery, prefilled syringe, auto-injector, pump, or other physical means of delivering the drug to the body.
User	The individual who incorporates that product or service into their life. These can be on-line banking customers, riders on a train, or drivers of a car. In our book, we take that to mean the patient who receives the medication or uses the medical device (combination product) to take their medication. It can also be a family member/caregiver who supports the patient in their health care.

MIX
Papier aus verantwortungsvollen Quellen
Paper from responsible sources
FSC® C105338

If you have any concerns about our products,
you can contact us on
ProductSafety@springernature.com

In case Publisher is established outside the EU,
the EU authorized representative is:
Springer Nature Customer Service Center GmbH
Europaplatz 3, 69115 Heidelberg, Germany

Printed by Libri Plureos GmbH
in Hamburg, Germany